OPPOSING VIEWPOINTS® SERIES

Renewable and Alternative Energy

Other Books of Related Interest

Opposing Viewpoints Series

America's Infrastructure and the Green Economy
Green Politics
The Politics of Climate

At Issue Series

Environmental Racism and Classism
Pipelines and Politics
The Role of Science in Public Policy

Current Controversies Series

The Economics of Clean Energy
Fossil Fuel Industries and the Green Economy
Sustainable Consumption

> "Congress shall make no law … abridging the freedom of speech, or of the press."
>
> *First Amendment to the U.S. Constitution*

The basic foundation of our democracy is the First Amendment guarantee of freedom of expression. The Opposing Viewpoints series is dedicated to the concept of this basic freedom and the idea that it is more important to practice it than to enshrine it.

OPPOSING VIEWPOINTS® SERIES

Renewable and Alternative Energy

Lisa Idzikowski and Avery Elizabeth Hurt, Book Editors

GREENHAVEN PUBLISHING

Published in 2024 by Greenhaven Publishing, LLC
2544 Clinton Street,
Buffalo NY 14224

First Edition

Articles in Greenhaven Publishing anthologies are often edited for length to meet page requirements. In addition, original titles of these works are changed to clearly present the main thesis and to explicitly indicate the author's opinion. Every effort is made to ensure that Greenhaven Publishing accurately reflects the original intent of the authors. Every effort has been made to trace the owners of the copyrighted material.

Cover image: majeczka/Shutterstock.com

Library of Congress CataloginginPublication Data

Names: Idzikowski, Lisa, editor. | Hurt, Avery Elizabeth, editor.
Title: Renewable and alternative energy / edited by Lisa Idzikowski and Avery Elizabeth Hurt.
Description: First edition. | New York : Greenhaven Publishing, 2024. | Series:
Opposing viewpoints | Includes bibliographical references and index.
Identifiers: ISBN 9781534509658 (pbk.) | ISBN 9781534509665 (library bound)
Subjects: LCSH: Renewable energy sources--Juvenile literature.
Classification: LCC TJ808.R415 2024 | DDC 333.79'4--dc23

Manufactured in the United States of America

Website: http://greenhavenpublishing.com

Contents

Chapter 4: Can Renewable and Alternative Sources Produce Enough Energy for the Future?

The Importance of Opposing Viewpoints

Perhaps every generation experiences a period in time in which the populace seems especially polarized, starkly divided on the important issues of the day and gravitating toward the far ends of the political spectrum and away from a consensus-facilitating middle ground. The world that today's students are growing up in and that they will soon enter into as active and engaged citizens is deeply fragmented in just this way. Issues relating to terrorism, immigration, women's rights, minority rights, race relations, health care, taxation, wealth and poverty, the environment, policing, military intervention, the proper role of government—in some ways, perennial issues that are freshly and uniquely urgent and vital with each new generation—are currently roiling the world.

If we are to foster a knowledgeable, responsible, active, and engaged citizenry among today's youth, we must provide them with the intellectual, interpretive, and critical-thinking tools and experience necessary to make sense of the world around them and of the all-important debates and arguments that inform it. After all, the outcome of these debates will in large measure determine the future course, prospects, and outcomes of the world and its peoples, particularly its youth. If they are to become successful members of society and productive and informed citizens, students need to learn how to evaluate the strengths and weaknesses of someone else's arguments, how to sift fact from opinion and fallacy, and how to test the relative merits and validity of their own opinions against the known facts and the best possible available information. The landmark series Opposing Viewpoints has been providing students with just such critical-thinking skills and exposure to the debates surrounding society's most urgent contemporary issues for many years, and it continues to serve this essential role with undiminished commitment, care, and rigor.

The key to the series's success in achieving its goal of sharpening students' critical-thinking and analytic skills resides in its title—

Opposing Viewpoints. In every intriguing, compelling, and engaging volume of this series, readers are presented with the widest possible spectrum of distinct viewpoints, expert opinions, and informed argumentation and commentary, supplied by some of today's leading academics, thinkers, analysts, politicians, policy makers, economists, activists, change agents, and advocates. Every opinion and argument anthologized here is presented objectively and accorded respect. There is no editorializing in any introductory text or in the arrangement and order of the pieces. No piece is included as a "straw man," an easy ideological target for cheap point-scoring. As wide and inclusive a range of viewpoints as possible is offered, with no privileging of one particular political ideology or cultural perspective over another. It is left to each individual reader to evaluate the relative merits of each argument—as they see it, and with the use of ever-growing critical-thinking skills—and grapple with their own assumptions, beliefs, and perspectives to determine how convincing or successful any given argument is and how the reader's own stance on the issue may be modified or altered in response to it.

This process is facilitated and supported by volume, chapter, and selection introductions that provide readers with the essential context they need to begin engaging with the spotlighted issues, with the debates surrounding them, and with their own perhaps shifting or nascent opinions on them. In addition, guided reading and discussion questions encourage readers to determine the authors' point of view and purpose, interrogate and analyze the various arguments and their rhetoric and structure, evaluate the arguments' strengths and weaknesses, test their claims against available facts and evidence, judge the validity of the reasoning, and bring into clearer, sharper focus the reader's own beliefs and conclusions and how they may differ from or align with those in the collection or those of their classmates.

Research has shown that reading comprehension skills improve dramatically when students are provided with compelling, intriguing, and relevant "discussable" texts. The subject matter of these collections could not be more compelling, intriguing, or urgently relevant to today's students and the world they are poised to inherit. The anthologized articles and the reading and discussion questions that are included with them also provide the basis for stimulating, lively, and passionate classroom debates. Students who are compelled to anticipate objections to their own argument and identify the flaws in those of an opponent read more carefully, think more critically, and steep themselves in relevant context, facts, and information more thoroughly. In short, using discussable text of the kind provided by every single volume in the Opposing Viewpoints series encourages close reading, facilitates reading comprehension, fosters research, strengthens critical thinking, and greatly enlivens and energizes classroom discussion and participation. The entire learning process is deepened, extended, and strengthened.

For all of these reasons, Opposing Viewpoints continues to be exactly the right resource at exactly the right time—when we most need to provide readers with the critical-thinking tools and skills that will not only serve them well in school but also in their careers and their daily lives as decision-making family members, community members, and citizens. This series encourages respectful engagement with and analysis of opposing viewpoints and fosters a resulting increase in the strength and rigor of one's own opinions and stances. As such, it helps make readers "future ready," and that readiness will pay rich dividends for the readers themselves, for the citizenry, for our society, and for the world at large.

Introduction

> *"It's impossible to take effective action against global warming without addressing our impact on the rest of the living world, and vice versa."*
>
> *-Christian de Perthuis and Édouard Civel, French economists*

In 1988, the United Nations Environment Programme (UNEP) and the World Meteorological Organization (WMO) created the Intergovernmental Panel on Climate Change (IPCC). The IPCC is an independent group of experts from over a hundred nations and many fields of science. These experts carefully evaluate and explain the science on climate change. The purpose of the organization is to provide policymakers and world leaders with reliable information so they can make wise decisions about climate-related policy. In 1990, the group began issuing periodic reports assessing the current state of the science on climate change and the socioeconomic effects of a warming Earth. These reports have been increasingly disturbing.

A simple graph of average global temperatures since records have been kept clearly shows the trend. Since the mid-1800s, when the Industrial Revolution began, global temperatures have skyrocketed. The reason? The burning of fossil fuels. In order to power factories, heat homes, and fuel trains, airplanes, cars, and trucks, humans began burning large amounts of fossil fuels, such as coal, oil, and gas. And the amounts have only increased as the world population has grown and countries have become more industrialized. Burning fossil fuels creates greenhouse gas emissions. And these gasses trap the sun's heat inside the atmosphere, making Earth hotter.

None of this is news, and there is no serious debate about any of it within the scientific community. What there is debate about, however, is how to solve the problem. It's clear that if we're going to prevent the worst consequences from climate change (many disastrous effects have already begun), we must drastically reduce or even eliminate the use of fossil fuels. And the way to do that is by using alternative and renewable forms of energy to power our civilization. These include solar and wind power, and maybe even nuclear power. But details about the best way to incorporate these energy sources and the economic and social consequences of implementing them are complex. In this volume, authors from many fields and across the political spectrum address these issues.

In the first chapter, the authors take a look at a very big question: Can renewable and alternative energy be implemented on a global scale? The problem won't be solved by anything less than a comprehensive worldwide effort. The chapter opens with a viewpoint that explains what renewable energy is and gives examples. Subsequent viewpoints look at how different countries are stepping up to the challenge.

If the countries that emit the most greenhouse gases shift to renewable and alternative energy, will that be enough to prevent a climate crisis? That's what the authors in Chapter 2 discuss. Some say that renewables are the answer—and the only answer—to this enormous problem. Others say that while they're a necessary part of the solution, they are not enough.

The authors in Chapter 3 approach the issue from a practical perspective. Is making the shift to renewable and alternative energy sources even feasible? Viewpoints here look at both the current successes and the possible challenges in the efforts to replace fossil fuels with more sustainable energy sources.

The world's population is growing, and the way we live is always changing. That means that when making important decisions like these, we have to consider the future as well as the present. The final chapter in this volume asks the hard question of whether renewable and alternative sources of energy can meet the world's needs in the future. The authors here take different approaches to the question but are largely optimistic. Some authors even argue that shifting to renewables would not only prevent climate apocalypse, but would improve the health and the economic situation of people worldwide.

Climate change is a huge problem. Renewable and alternative sources of energy are an obvious response. In *Opposing Viewpoints: Renewable and Alternative Energy*, the authors offer important details and much-needed nuance to the conversation.

OPPOSING VIEWPOINTS® SERIES

CHAPTER 1

Can Renewable and Alternative Energy Be Implemented on a Global Scale?

Chapter Preface

If renewable and alternative energy are going to make a difference in the future of the planet, the alternative forms of energy must be widely used. One country here and another there making changes, even big changes, will not be enough. The effort must be global. In this chapter, the authors discuss many of the issues involved in implementing renewable energy around the globe, particularly in the nations with the largest populations: India, China, and the United States. (In 2023, India passed China in terms of population.)

The United States is the second largest emitter of carbon dioxide after China. One of the authors here looks at efforts in the U.S. to shift to renewable sources of energy. He explains that use of renewable energy is growing rapidly, but fossil fuels still make up the lion's share of energy use in the U.S. The next viewpoint takes a close look at how the world might go about meeting the United Nations' Sustainable Development Goals. If done right, the authors argue, this will not do economic harm, as many have suggested, but might actually improve the economies of nations that step up to the task.

Two viewpoints focus on the other two countries that are major sources of emissions: India and China. The viewpoint about India looks at that nation's goal to eliminate the production of all greenhouse gas emissions by 2070. There are challenges to meeting this goal, and the author notes that their short-term goals are considerably less ambitious, but the country's bold vision could benefit the environment and the economy. The viewpoint about China looks at how China's energy use has changed in recent years and the role the country has played in helping other countries turn to renewable energy, specifically in Latin America. In recent years, the nation that is by far the world's largest emitter of carbon dioxide has also become the leader in solar and wind power. The reasons China has moved toward renewables, the authors explain, have as much to do with geopolitics and economics as environmental protection. This relationship between China and other countries has raised some political concerns, but it has played an important role in promoting the renewable energy industry around the world.

Viewpoint 1

> *"Because of the energy crisis in the United States during the 1970s, dwindling supplies of fossil fuels and hazards associated with nuclear power, usage of renewable energy sources such as solar energy, hydroelectric, wind, biomass, and geothermal has grown."*

There Is an Urgent Need for Renewable Energy

Jean Brainard

In this excerpted viewpoint, Jean Brainard explains why it is essential for industrialized nations to become less dependent on fossil fuels and turn to other sources of energy, specifically renewable sources. She also points out why this is so challenging to accomplish. She suggests that one of the main challenges impacting all types of renewable energy is that the current energy infrastructure was created around fossil fuels, so switching to renewables would require a complicated and expensive overhaul of the infrastructure. She also discusses the different types of renewable energy, their advantages, and any challenges associated with them. Although the shift to renewable energy may not be easy, she asserts that the limited amount of fossil fuels available makes it a necessity. Jean Brainard is a freelance science and health writer and curriculum specialist.

As you read, consider the following questions:

1. According to this viewpoint, how long will the remaining reserves of coal, gas, and oil last us?
2. Why isn't nuclear energy considered a renewable energy?
3. How is most hydroelectric power generated?

Sufficient, reliable sources of energy are a necessity for industrialized nations. Energy is used for heating, cooking, transportation, and manufacturing. Energy can be generally classified as nonrenewable and renewable. Over 85 percent of the energy used in the world is from nonrenewable supplies. Most developed nations are dependent on nonrenewable energy sources such as fossil fuels (coal and oil) and nuclear power. These sources are called nonrenewable because they cannot be renewed or regenerated quickly enough to keep pace with their use. Some sources of energy are renewable or potentially renewable. Examples of renewable energy sources are: solar, geothermal, hydroelectric, biomass, and wind. Renewable energy sources are more commonly used in developing nations.

A nonrenewable resource is not regenerated or restored on a time scale comparative to its consumption. Nonrenewable resources exist in fixed amounts (at least relative to our time frame), and can be used up. The classic examples are fossil fuels such as petroleum, coal, and natural gas. Fossil fuels have formed from remains of plants (for coal) and phyto- and zooplankton (for oil) over periods from 50 to 350 million years. Ecologist Jeff Dukes estimates that 20 metric tons of phytoplankton produce 1 liter of gasoline! We have been consuming fossil fuels for less than 200 years, yet even the most optimistic estimates suggest that remaining reserves can supply our needs for

- Coal: 252 years
- Gas: 72 years
- Oil: 45 years

Nuclear power is considered a nonrenewable resource because uranium fuel supplies are finite. Some estimates suggest that known economically feasible supplies could last 70 years at current rates of use—although known, and probably unknown reserves are much larger, and new technologies could make some reserves more useful.

Global energy use includes mostly nonrenewable (oil, coal, gas, and nuclear) but increasing amounts of renewable (biomass, hydro, solar, wind, geothermal, biofuels, and solar photovoltaic) resources.

Today 80 percent of the world's energy comes from fossil fuels. Fossil fuels are formed hundreds of millions of years ago from decomposition of the bodies of living things became buried.

The reasons why fossil fuels are so widely used today:

1. Fossil fuel is easy to store and transport.
2. Society's energy infrastructure and technology for transportation is geared for using fossil fuel energy.
3. Fossil fuels like coal and oil are easy to store and transport, while natural gas is used in countries that have natural gas pipeline networks.

Society's dependence on fossil fuel is problematic because:

1. Fossil fuel supply are finite and crude oil supplies are running low.
2. Most crude oil producing countries are developing and politically unstable, so that crude oil can suddenly be in short supply due to politics. E.g. OPEC's (Organization of Petroleum Exporting Countries) 1973-74 oil embargo against the United States for supporting Israel in the Arab-Israel war caused widespread panic, skyrocketing prices, and spurred inflation.
3. Burning fossil fuels creates air pollution and causes acid rain.
4. Burning fossil fuels released carbon dioxide into the atmosphere and contributes to global climate change.

[…]

Renewable Energy

Renewable energy sources are often considered alternative sources because, in general, most industrialized countries do not rely on them as their main energy source. Instead, they tend to rely on nonrenewable sources such as fossil fuels or nuclear power. Because of the energy crisis in the United States during the 1970s, dwindling supplies of fossil fuels and hazards associated with nuclear power, usage of renewable energy sources such as solar energy, hydroelectric, wind, biomass, and geothermal has grown.

Renewable energy comes from the sun (considered an "unlimited" supply) or other sources that can theoretically be renewed at least as quickly as they are consumed. If used at a sustainable rate, these sources will be available for consumption for thousands of years or longer. Unfortunately, some potentially renewable energy sources, such as biomass and geothermal, are actually being depleted in some areas because the usage rate exceeds the renewal rate.

A resource replenished by natural processes at a rate roughly equal to the rate at which humans consume it is a renewable resource. Sunlight and wind, for example, are in no danger of being used in excess of their longterm availability. Hydropower is renewed by Earth's hydrologic cycle. Water has also been considered renewable, but overpumping of groundwater is depleting aquifers, and pollution threatens the use of many water resources, showing that the consequences of resource use are not always simple depletion. Soils are often considered renewable, but erosion and depletion of minerals proves otherwise. Living things (forests and fish, for example) are considered renewable because they can reproduce to replace individuals lost to human consumption. This is true only up to a point, however; overexploitation can lead to extinction, and overharvesting can remove nutrients so that soil fertility does not allow forest renewal. Energy resources derived from living things, such as ethanol, plant oils, and methane, are considered renewable, although their costs to the environment are not always adequately considered. Renewable materials would include sustainably harvested wood, cork, and bamboo as well as sustainably harvested crops. Metals and other minerals are sometimes considered renewable because they are not destroyed when they are used and can be recycled.

Although most of our energy needs today come from fossil fuels but we are now increasingly aware that we need to find alternative energy sources. The benefits of using renewable energy sources are:

1. Most energy will never run out.
2. Countries are no longer so dependent on oil-producing nationals for their energy needs.
3. Most renewable energy sources produce little or no air pollution.
4. Most renewable energy sources produce little or no carbon dioxide into the atmosphere and therefore is a solution to global climate change.

The reasons why renewable energy sources are NOT more widely used:

1. Most of our technology and infrastructure today is geared toward using fossil fuel energy.
2. Switching to renewable energy requires expensive investments in infrastructure.
3. We have limited ways to transport electricity generated by renewable energy sources that are not near electric grids.

The various ways we produce electricity from renewable energy sources is based on the principle that something has to turn turbines to generate electricity. The turning force spins giant magnets suspended within coils of metal wire inside the turbine which creates electricity inside the metal coils.

Solar Energy

Solar energy is the ultimate energy source driving the earth. Though only one billionth of the energy that leaves the sun actually reaches the earth's surface, this is more than enough to meet the world's energy requirements. In fact, all other sources of energy, renewable and nonrenewable, are actually stored forms of solar energy. The process of directly converting solar energy to heat or electricity is considered a renewable energy source. Solar energy represents an essentially unlimited supply of energy as the sun will long outlast human civilization on earth. The difficulties lie in harnessing the energy. Solar energy has been used for centuries to heat homes and water, and modern technology (photovoltaic cells) has provided a way to produce electricity from sunlight.

There are two basic forms of radiant solar energy use: passive and active. Passive solar energy systems are static, and do not require the input of energy in the form of moving parts or pumping fluids to utilize the sun's energy.

Buildings can be designed to capture and collect the sun's energy directly. Materials are selected for their special characteristics: glass allows the sun to enter the building to provide light and heat; water and stone materials have high heat capacities. They can absorb large amounts of solar energy during the day, which can then be used during the night. A southern exposure greenhouse with glass windows and a concrete floor is an example of a passive solar heating system. Active solar energy systems require the input of some energy to drive mechanical devices (e.g., solar panels), which collect the energy and pump fluids used to store and distribute the energy. Solar panels are generally mounted on a south or west-facing roof. A solar panel usually consists of a glass-faced, sealed, insulated box with a black matte interior finish. Inside are coils full of a heat-collecting liquid medium (usually water, sometimes augmented by antifreeze).

The sun heats the water in the coils, which is pumped to coils in a heat transfer tank containing water. The water in the tank is heated and then either stored or pumped through the building to heat rooms or supply hot water to taps in the building.

It has a long history of use by preindustrial age human society, e.g. using the sun to dry food or clothes. There are two ways to harness solar energy.

1. Harness the sun's heat (either directly or by concentrating the heat)
2. Convert the sun's light to electricity with photovoltaic cells

Photovoltaic cells generate electricity from sunlight. Hundreds of cells are linked together to provide the required flow of current. The electricity can be used directly or stored in storage batteries. Because photovoltaic cells have no moving parts, they are clean, quiet, and durable. Early photovoltaic cells were extremely expensive, making the cost of solar electric panels prohibitive. The recent development of inexpensive semiconductor materials has helped greatly lower the cost to the point where solar electric panels can compete much better cost-wise with traditionally produced electricity.

Though solar energy itself is free, large costs can be associated with the equipment. The building costs for a house heated by passive solar energy may initially be more expensive. The glass, stone materials, and excellent insulation necessary for the system to work properly tend to be more costly than conventional building materials. A long-term comparison of utility bills, though, generally reveals noticeable savings. The solar panels used in active solar energy can be expensive to purchase, install and maintain. Leaks can occur in the extensive network of pipes required, thereby causing additional expense. The biggest drawback of any solar energy system is that it requires a consistent supply of sunlight to work. Most parts of the world have less than ideal conditions for a solar-only home because of their latitude or climate. Therefore, it is usually necessary for solar houses to have conventional backup systems (e.g. a gas furnace or hot-water heater). This double-system requirement further adds to its cost.

Wind Power

Wind is the result of the sun's uneven heating of the atmosphere. Warm air expands and rises, and cool air contracts and sinks. This movement of the air is called wind. Wind has been used as an energy source for millennia. It has been used to pump water, to power ships, and to mill grains. Areas with constant and strong winds can be used by wind turbines to generate electricity. In the United States, the state of California has about 20,000 wind turbines and produces the most wind-generated electricity. Wind energy does not produce air pollution, can be virtually limitless, and is relatively inexpensive to produce. There is an initial cost of manufacturing the wind turbine and the costs associated with upkeep and repairs, but the wind itself is free.

The major drawbacks of wind-powered generators are that they require lots of open land and a fairly constant wind supply. Less than 15 percent of the United States is suitable for generating wind energy. Windmills are also noisy, and some people consider them aesthetically unappealing and label them as visual pollution. Migrating birds and insects can become entangled and killed by the turning blades. However, the land used for windmill farms can be simultaneously used for other purposes such as ranching, farming, and recreation.

1. Has a long history of use by preindustrial age human society, e.g. windmills used to grind grain into flour
2. The fastest growing source of renewable energy

Wind power is considered a renewable resource because the rate of supply far exceeds the rate of use.

Hydroelectric Power

Hydroelectric power is generated by using the energy of flowing water to power generating turbines for producing electricity. Most hydroelectric power is generated by dams across large-flow rivers. A dam built across river creates a reservoir behind it. The height of the water behind the dam is greater than that below the dam, representing stored potential energy. When water flows down through the penstock of the dam, driving the turbines, some of this potential energy is converted into electricity. Hydroelectric power, like other alternative sources, is clean and relatively cheap over the long term even with initial construction costs and upkeep. But because the river's normal flow rate is reduced by the dam, sediments normally carried downstream by the water are instead deposited in the reservoir. Eventually, the sediment can clog the penstocks and render the dam useless for power generation.

Large-scale dams can have a significant impact on the regional environment. When the river is initially dammed, farmlands are sometimes flooded and entire populations of people and wildlife are displaced by the rising waters behind the dam. In some cases, the reservoir can flood hundreds or thousands of square miles or kilometers. The decreased flow downstream from the dam can also negatively impact human and wildlife populations living downstream. In addition, the dam can act as a barrier to fish that must travel upstream to spawn. Aquatic organisms are frequently caught and killed in the penstock and the out-take pipes. Because of the large surface area of the reservoir, the local climate can change due to the large amount of evaporation occurring.

- Has a long history of use by preindustrial age human society, e.g. watermills used to grind grain in to flour.
- Today hydropower mostly involves building dams across major rivers and harnessing the power of the water as it flows across the dam to turn turbines that generate electricity.
- The world has maximized the use of hydropower because most of the world's major rivers have dams on them.
- Building of dams causes major habitat destruction due to the large area behind the dams that become flooded.
- This energy is renewable only if the river continues to flow.

Biomass Energy

Biomass energy is the oldest energy source used by humans. Biomass is the organic matter that composes the tissues of plants and animals. Until the Industrial Revolution prompted a shift to fossil fuels in the mid 18th century, it was the world's dominant fuel source. Biomass can be burned for heating and cooking, and even generating electricity. The most common source of biomass energy is from the burning of wood, but energy can also be generated by burning animal manure (dung), herbaceous plant material (non-wood), peat (partially decomposed plant and animal tissues), or converted biomass such as charcoal (wood that has been partially burned to produce a coal-like substance). Biomass can also be converted into a liquid biofuel such as ethanol or methanol. Currently, about 15 percent of the world's energy comes from biomass.

Biomass is a potentially renewable energy source. Unfortunately, trees that are cut for firewood are frequently not replanted. In order to be used sustainably, one tree must be planted for every one cut down.

Biomass is most frequently used as a fuel source in developing nations, but with the decline of fossil fuel availability and the increase in fossil fuel prices, biomass is increasingly being used as a fuel source in developed nations. One example of biomass energy in developed nations is the burning of municipal solid waste. In the United States, several plants have been constructed to burn urban biomass waste and use the energy to generate electricity.

The use of biomass as a fuel source has serious environmental effects. When harvested trees are not replanted, soil erosion can occur. The loss of photosynthetic activity results in increased amounts of carbon dioxide in the atmosphere and can contribute to global warming. The burning of biomass also produces carbon dioxide and deprives the soil of nutrients it normally would have received from the decomposition of the organic matter. Burning releases particulate matter (such as ash) into the air which can cause respiratory health problems.

- Biomass: organic material that makes up living organisms.

  - More than 1 billion people use wood from trees as their principal energy source.
  - In developing nations, families gather fuelwood for heating, cooking, and lighting.
  - Biomass is only renewable if not overharvested.
  - Cofiring combines biomass with coal to burn and generate electricity.
- Gasification: vaporized biomass at high temperature without oxygen produce a gas mix that can be used to produce electricity, methanol, and diesel fuel.
- Biofuels: biomass sources are converted into fuels to power automobiles.
- Ethanol: produced as a biofuel by fermenting plant matter (e.g. carbohydrate-rich crops, crop residue, weeds).
  - Ethanol is widely added to U.S. gasoline to reduce emissions.
  - Any vehicle on the road today will run well on a 10 percent ethanol mix.
  - But ethanol made from corn can cause food prices to increase, therefore it is better to use waste plant materials for making ethanol.
- Flexible fuel vehicles: run on 85 percent ethanol.
- Biodiesel: a fuel produced from vegetable oil, used cooking grease or animal fat used in diesel engines.

Geothermal Energy

Geothermal energy uses heat from the earth's internal geologic processes in order to produce electricity or provide heating. One source of geothermal energy is steam. Groundwater percolates down though cracks in the subsurface rocks until it reaches rocks heated by underlying magma, and the heat converts the water to steam. Sometimes this steam makes its way back to the surface in the form of a geyser or hot spring. Wells can be dug to tap the steam reservoir and bring it to the surface, to drive generating turbines and produce electricity. Hot water can be circulated to heat buildings. Regions near tectonic plate boundaries have the best potential for geothermal activity.

The western portion of the United States is most conducive for geothermal energy sources, and over half of the electricity used by the city of San Francisco comes from the Geysers, a natural geothermal field in Northern California. California produces about 50 percent of the world's electricity that comes from geothermal sources.

Entire cities in Iceland, which is located in a volcanically active region near a mid-ocean ridge, are heated by geothermal energy. The Rift Valley region of East Africa also has geothermal power plants. Geothermal energy may not always be renewable in a particular region if the steam is withdrawn at a rate faster than it can be replenished, or if the heating source cools off. The energy produced by the Geysers region of California is already in decline because the heavy use is causing the underground heat source to cool. Geothermal energy recovery can be less environmentally invasive than engaging in recovery methods for nonrenewable energy sources. Although it is relatively environmentally friendly, it is not practical for all situations. Only limited geographic regions are capable of producing geothermal energy that is economically viable. Therefore, it will probably never become a major source of energy. The cost and energy requirements for tapping and transporting steam and hot water are high. Hydrogen sulfide, a toxic air pollutant that smells like rotten eggs, is also often associated with geothermal activity.

There are two ways to harness geothermal heat:

- Geothermal heat pumps: Utilize the temperature difference between above ground and below ground temperatures.
    - Underground temperature stays constant year round (45-75°F).

- Antifreeze solution circulates from the house/building to the ground.
- The pumps heat buildings in the winter by transferring heat from the ground into buildings.
- In the summer, heat is transferred through underground pipes from the house/building into the ground.

- Geothermal power stations: Harness energy for electricity productions by using hot water heated to high temperatures in the ground in places where hot magma is closer to Earth's crust (e.g. hot springs or geysers).
  - Geothermal power plants use naturally heated water and steam for direct heating or generating electricity.

Viewpoint 2

> *"Solar and wind power use has grown at a rapid rate over the past decade or so, but as of 2018 those sources accounted for less than 4 percent of all the energy used in the U.S."*

Renewable Energy Is Growing Fast, but Fossil Fuels Still Dominate

Drew DeSilver

In this viewpoint, which was originally published in 2020, Drew DeSilver points out that 77 percent of Americans consider it important to develop alternative sources of energy. Then he takes a careful look at where the nation's energy currently comes from and how much that has changed over the last few decades. He points out that although there has been considerable growth in renewable energy usage, fossil fuels are still the most commonly used source of energy by a significant margin. Drew DeSilver is a senior writer and editor at the Pew Research Center.

As you read, consider the following questions:

1. According to this viewpoint, what is the source of the majority of the United States' energy?
2. How has U.S. energy usage changed in recent years?

"Renewable energy is growing fast in the U.S., but fossil fuels still dominate," by Drew DeSilver, Pew Research Center, January 15, 2020.

3. What was the biggest renewable energy source as of 2018, the last year data was available as of the publication of this viewpoint?

Most Americans (77 percent) say it's more important for the United States to develop alternative energy sources, such as solar and wind power, than to produce more coal, oil, and other fossil fuels, according to a recent Pew Research Center survey. Which raises the question: How *does* the U.S. meet its vast energy needs, and how, if at all, has that changed?

The answer, as one might expect, is complicated. Solar and wind power use has grown at a rapid rate over the past decade or so, but as of 2018 those sources accounted for less than 4 percent of all the energy used in the U.S. (That's the most recent full year for which data is available.) As far back as we have data, most of the energy used in the U.S. has come from coal, oil, and natural gas. In 2018, those "fossil fuels" fed about 80 percent of the nation's energy demand, down slightly from 84 percent a decade earlier. Although coal use has declined in recent years, natural gas use has soared, while oil's share of the nation's energy tab has fluctuated between 35 percent and 40 percent.

The total amount of energy used in the U.S.—everything from lighting and heating homes to cooking meals, fueling factories, driving cars, and powering smartphones—hit 101.2 quadrillion Btu in 2018, the highest level since data collection began in 1949, according to the federal Energy Information Administration (EIA).

(Short for British thermal unit, Btu is often used in the energy industry—not to mention the home-appliance business—as a common yardstick to measure and compare different types of energy. One Btu is the amount of energy needed to heat 1 pound of water by 1 degree Fahrenheit at sea level. It's equivalent to about 1,055 joules in the metric system, or the heat released by burning a common wooden kitchen match.)

About 38 perent of all those Btu flowed into the electric power industry (electric utilities and independent power producers), which converted them into electricity and sent them back out into the rest of the economy. Transportation accounted for about 28 percent of total energy use, followed by the industrial sector (23 percent), households (7 percent), and commercial establishments (less than 5 percent).

Per capita energy use in the U.S. had been trending lower since the turn of the 21st century but ticked up in 2018. On average, each American in 2000 used about 349.8 million Btu. By 2017 that had fallen to 300.5 million Btu, the lowest level in five decades. In 2018, though, per capita energy use rose to 309.3 million Btu. (Per capita energy use peaked in 1979 at 359 million Btu.)

Looked at a different way, the U.S. economy has become steadily less energy-intensive since the end of World War II. In 1949, it took 15,175 Btu to generate each dollar of real gross domestic product. By 2018, it took 5,450, a 64 percent decrease. But there's still plenty of inefficiency in the system: The Lawrence Livermore National Laboratory calculated that in 2018, about two-thirds of all the energy used was wasted (as with heat exhaust from vehicles and furnaces). And only 34.5 percent of the energy used by the electric power industry reaches end users as electricity—the rest is lost in the process of generating, transmitting, and distributing the power.

Oil and Natural Gas Production Up, Coal Down

Today, the United States meets nearly all its energy needs through domestic production. Net imports, mainly petroleum, accounted for less than 4 percent of the total U.S. energy supply in 2018, versus 26 percent a decade earlier.

In the first 10 months of 2019, the U.S. pumped nearly 3.7 billion barrels of crude oil, more than 2 billion more than in the same period in 2009, according to EIA data. For the full year 2018, crude accounted for nearly a quarter of all U.S. energy production. Natural gas, which accounted for about a third of total energy production in 2018, has also soared—from 21.7 trillion cubic feet in the first nine months of 2009 to 33.6 trillion cubic feet in the same period in 2019.

Those dramatic increases in domestic oil and gas production have been driven by new technologies, most notably fracking

What Is Green Energy?

What is green energy? And what's the difference between green energy, renewable energy, and clean energy? Often these terms are used interchangeably, but there are some differences.

One thing that green, clean, and renewable energy all have in common is that they're being increasingly used to generate electricity in order to phase out the use of fossil fuels, like coal and gas, which are a key cause of climate change.

What Is the Definition of Green Energy?

Green energy is energy that can be produced using a method, and from a source, that causes no harm to the natural environment.

What's the Difference Between Green Energy and Renewable Energy?

The terms '"green energy" and "renewable energy" are often used interchangeably, but there is one essential (and sometimes confusing) difference between them. While most green energy sources are also renewable, not all renewable energy sources are considered entirely green.

and horizontal drilling, which enable companies to access underground deposits that previously were too expensive to tap. As a consequence, the U.S. was the world's biggest producer of both oil and gas in 2018—ahead of Saudi Arabia and Russia, respectively.

Coal, on the other hand, has declined precipitously since its 2008 peak, when nearly 1.2 billion tons were mined. Nearly all U.S. coal (about 93 percent in 2018, according to EIA data) is used to generate electricity. But as a Brookings Institution report notes, U.S. electricity demand has been stagnant, the price of natural gas has fallen as production has soared, and government policy has until recently favored other energy sources such as wind and solar. In 2018, coal accounted for just 16 percent of total domestic energy production, less than half its share a decade earlier. The amount mined in the first nine months of 2019, 540 million tons, was about a third lower than in the same period in 2009.

Renewable energy comes from sources that are constantly and naturally renewed (hence the name), such as wind power and solar power. Renewable energy is also often called sustainable energy.

A renewable energy source may not be considered "green" if, for example, some carbon emissions are associated with the processes used to generate the energy—such as the building of infrastructure.

What's the Differences Between Green Energy and Clean Energy?

Clean energy is energy that, when used, creates little or no greenhouse gas emissions.

As with renewable energy, some types of clean energy may not always be considered entirely green.

Here's an easy way to differentiate between clean energy, green energy and renewable energy:

- Clean energy = clean air
- Green energy = no harm to the environment
- Renewable energy = sources that replenish naturally, such as the sun and the wind

"What is green energy?," National Grid.

Over the past decade, solar power has experienced the largest percentage growth of any U.S. energy source. Solar generated just over 2 billion kilowatt-hours of electricity in 2008. A decade later, it generated more than 93 billion kilowatt hours, an almost 46-fold increase. Solar's growth is occurring on both the large scale (electric power plants) and the small scale (rooftop solar panels). Overall, about two-thirds of all solar energy was produced by electric utilities, with solar setups on homes and commercial buildings accounting for most of the rest.

Still, solar accounted for only 1 percent of the nation's total energy production in 2018. The biggest renewable energy source remained hydropower (2.8 percent of total production), followed by wind, wood, and biofuels.

Viewpoint 3

> *"Contrary to some commonly held beliefs, a clean transition is, and has been for at least a decade, good for the economy—even in earlier stages of its development."*

Transitioning to Renewable Energy Can Benefit the Environment and the Economy

Deborah de Lange

In this viewpoint, Deborah de Lange argues that it is not necessary to choose between economic prosperity and sustainable energy—investing in renewable energy sources allows for both to take place. She also argues that rather than continuing to subsidize the fossil fuel industry—which only benefits those within the industry, not consumers—governments should start implementing pollution taxes. This can provide money to support renewable energy research and innovation. Through looking at the issue of consumer welfare through a different lens, it becomes clear that the renewable energy industry offers more benefits than the fossil fuel industry. Deborah de Lange is an associate professor of global management studies at Toronto Metropolitan Universities.

"Renewable energy innovation isn't just good for the climate — it's also good for the economy," by Deborah de Lange, The Conversation, February 26, 2024, https://theconversation.com/renewable-energy-innovation-isnt-just-good-for-the-climate-its-also-good-for-the-economy-223164.

As you read, consider the following questions:

1. Why does de Lange say that fossil fuels are bad for consumer welfare despite raising GDP?
2. What does de Lange think should be taken into consideration when measuring economic prosperity beyond GDP?
3. What recommendations does de Lange make at the end of the viewpoint?

As the climate crisis escalates, there are urgent and difficult choices that need to be made to drastically reduce our carbon emissions before more irreparable damage is done.

Many have argued the energy industry needs to change to reduce carbon emissions, but one concern that remains is the consequence this will have on economic prosperity.

Discussions vary across interest groups. Do we need to outright replace the fossil fuel industry with the renewable energy industry as soon as possible? Should we slowly phase out fossil fuels while making clean renewable replacements? Or, should we continue with a powerful fossil fuel industry while separately growing a renewable industry in parallel?

How these different choices could impact our economies seems unclear, and it is this lack of clarity that opens up the field for frustrating discussions. At the recent COP28 climate summit in the United Arab Emirates, the conference president shockingly said that there is "no science" behind any decision to phase-out fossil fuels from our energy systems—a statement which he later claimed was "misinterpreted."

My recent research examines energy industry restructuring options for a green transition to renewable energy from an economic perspective.

Although economic analysis is helpful, it is not sufficient on its own for making these important decisions. So, my research also draws on sustainability which involves considering the conditions faced by future generations, and a concept known as equifinality reminding us to keep our minds open to many possible approaches that may satisfy the same objectives.

Renewable Energy Innovation and GDP

My research indicates that renewable energy innovation contributes to higher GDP. Contrary to some commonly held beliefs, a clean transition is, and has been for at least a decade, good for the economy—even in earlier stages of its development.

My findings also suggest that government and industry support for the fossil fuel industry negatively affects a country's renewable energy innovation. The two industries are not compatible.

When the fossil fuel industry invests in itself, it also appears to improve GDP, which creates confusion about the best way to ensure economic prosperity while transitioning to clean energy.

But this investment, often made through lobbying, only prolongs the existence of the fossil fuel industry by keeping renewable energy competition out. This creates a false dichotomy between reducing emissions and improving GDP when, in fact, clean innovation can achieve both simultaneously.

My research indicates that clean innovation makes a stronger economy *and* reduces emissions. If we want to reinforce that dual progress, rather than accepting trade-offs, then we have to stop supporting the fossil fuel industry which aims to slow it down.

Helping Renewable Energy Thrive

Economically speaking, the fossil fuel industry is negatively impacting consumer welfare by maintaining higher-than-necessary prices due to limited competition. This, in turn, bumps up GDP through inflated profits, having subsidised an already dominant polluting industry, reducing clean innovation, and delaying cleaner progress—obviously not the way to grow a healthy economy.

In fact, GDP is not a standard of living measure or a measure of innovative competitiveness. To address inflation and the cost of living crisis, we should be promoting more competition across industries. This is a more productive type of capitalism that brings wider benefits to all of us, including more innovation, lower prices, and better products for domestic and export markets.

Government subsidies that boost the fossil fuel industry hinder consumer welfare and the transition to clean energy. Some examples include subsidies to fund more carbon capture and storage technology and the use of fossil energy in hydrogen storage systems.

Instead of funding these backward subsidies, governments should implement pollution taxes while also supporting renewable energy innovation. My research demonstrates that pollution taxes work well with clean innovation capabilities. Supporting research and innovation in renewable energy and using a carbon tax as a tool can boost the economic benefits of transitioning to clean energy.

The findings of my work suggest that a robust economy is related to industry restructuring so that renewable energy innovation can thrive. Fostering novel scientific discoveries in clean energy innovation should be prioritized while reducing non-competitive industry formations and organizations, such as fossil fuel oligopolies and industry associations.

Making Decisions with the Future in Mind

Increasing public awareness and understanding of fossil fuel industry games is a way to accelerate change. It's important to recognize that industries at different life cycle stages contribute to the economy in different ways.

A newer rising industry with determined entrepreneurs, like that of renewable energy, invests in innovation to create value. On the other hand, a declining industry plays end-game strategies, like engaging in self-promotional activities, to maintain their existing position and create barriers to new industry entries.

However, consumer welfare increases with competition, not collusion. Economic analysis is not sufficient on its own for decision-making in this area because positive economic outcomes can be generated by different kinds of strategies supporting an industry's life cycle goals.

Government policy decisions should be made based on economic analyses alongside broader sustainability criteria. Ignoring the equifinality argument and reverting to discussions about replacing coal with gas as a bridge only ensures fossil fuels remain in use for at least another generation of infrastructure.

Communities should apply sustainability and equifinality lenses to decision-making, understanding that there are many possible means to an end. For example, if a community has specific concerns about one type of renewable energy system, they should explore other alternative clean energy options instead of defaulting to fossil fuels.

An educated public should reject simplistic and single-sided arguments and understand there are usually more nuanced solutions to problems supported by evidence-based analysis. By embracing a more holistic approach, we can develop more sustainable societies by opening up renewable energy possibilities.

Viewpoint 4

> *"India has the potential to set the tone for emerging economies' climate action over the coming decade."*

India's UN Climate Goals Aim Low, but Its National Policies Are Ambitious

Tarun Gopalakrishnan

This viewpoint by Tarun Gopalakrishnan discusses research he completed with the Climate Policy Lab at Tufts University on India's climate policies. The research indicates that India has ambitious climate policies on a national level, but that it set modest goals during the UN climate talks in Glasgow in 2021. Gopalakrishnan argues that the commitments India submitted to the UN suggest a "business as usual" approach to emissions, but that it is taking drastic steps to increase the use of non-fossil electricity to power the country. The Indian Parliament as well as private companies are working to support these goals. It is uncertain whether India will be able to meet its bolder goals, but the fact that it is prioritizing a shift to cleaner energy is a move in the right direction that could help encourage other emerging economies to do the same. Tarun Gopalakrishnan is a junior research fellow in the Climate Policy Lab at Tufts University.

As you read, consider the following questions:

1. What targets for 2030 did India include in its 2022 climate commitments?
2. What changes that have occurred in electricity generation could help India meet its goals?
3. Why does Gopalakrishnan say that it could have benefitted India to share bolder goals before the UN?

At the United Nations climate talks in Glasgow in 2021, Indian Prime Minister Narendra Modi surprised the world when he announced that his country would zero out its greenhouse gas emissions by the year 2070. It was a landmark decision, acknowledging that long-term decarbonization is in India's interest.

However, climate change is threatening lives, crops and India's economy today. New Delhi endured extreme heat for several weeks in early 2022, with temperatures regularly crossing 104 degrees Fahrenheit (40 Celsius). The previous year, cyclones, flash floods and extreme rainfall destroyed more than 12 million acres of crops, contributing to a global spike in food prices. At the same time, energy demand is rising in a country forecast to pass China as the world's most populous in 2023.

So, when the dust settled around the net zero announcement, scrutiny turned to India's short-term ambitions for the coming decade.

On Aug. 26, 2022, India formally submitted its second set of international climate commitments, known as its Nationally Determined Contribution, or NDC, to the United Nations, including its short-term climate targets and strategies for meeting them.

India has the potential to set the tone for emerging economies' climate action over the coming decade. However, its NDC commitments significantly understate the ambition in its own national climate policies. These mixed signals could slow down

India's burgeoning energy transition and hamper its ability to raise international climate finance.

India's 2030 Climate Targets

India's new climate commitments include two primary targets for 2030. One is to reduce emissions per unit of gross domestic product, or GDP, by 45 percent, relative to the year 2005. The other is to increase "non-fossil" electricity—solar, wind, nuclear ,and hydropower—to half of the country's electricity capacity.

While these targets are an improvement over India's commitments when it joined the Paris climate agreement in 2015, they are largely a continuation of the country's "business-as-usual" emissions trajectory. A fast-growing country can reduce its emissions per GDP and increase its emissions.

Views differ on whether this is acceptable. There is considerable debate around what each country's "fair share" of the global carbon budget is, given industrialized countries' significantly larger contribution to per capita and cumulative greenhouse emissions.

India's State Climate Policies Are More Ambitious

India currently meets about a quarter of its electricity demand with nonfossil energy, about 160 gigawatts of power generation capacity in all. It added 15.4 gigawatts of solar and wind capacity in 2021–22, the third-highest increase in the world.

In its national policies, India has stated that it intends to more than triple nonfossil electricity capacity to 500 gigawatts by 2030.

That's an ambitious increase, but it draws inspiration from evolving realities: Electricity generated from renewables is now cheaper at auction than coal-fired power. Renewable energy with energy storage is also expected to be cheaper than coal within this decade, driven partly by the government's $2.5 billion boost for energy storage manufacturing in India.

Displacing coal as the grid's primary generation source seems technically and economically viable at last.

The counterpoint is that India has set and missed ambitious renewable energy targets before—it will fall short of its goal, set in 2010, of reaching 100 gigawatts of solar and 60 gigawatts of wind power capacity by 2022. This may partially explain the reluctance to formally commit to higher international targets.

The government is also still granting loans for new coal power plants and providing higher subsidies for coal than for renewable energy. These are legacies of its "all-of-the-above" energy policy driven by the continuing need to drastically improve affordable energy access. While the government is beginning to consider what a "just transition" away from coal would cost, it is planning for an increase in coal production for power generation and industrial processes over the coming decade.

Parliament Weights Legally Enforceable Targets

The Indian Parliament is about to elevate some of these policy targets to legally enforceable mandates.

The Energy Conservation Amendment, passed by the lower house, proposes renewable power purchase targets for industries, inclusion of large residential buildings in the energy conservation code, and energy consumption standards for vehicles and vessels. It also empowers the government to create a national carbon trading market.

The private sector is paying attention to these policy signals, the falling cost of producing clean electricity and transportation, and the projected rising energy demand as the country develops. India's largest conglomerates have set aggressive renewable electricity expansion targets. Indian automakers are competing to invest in EV manufacturing and pushing the government to speed up disbursement of subsidies and incentives. But the same conglomerates are also expanding coal investments, illustrating the importance of avoiding mixed policy signals.

Positive Effects on Economies and Employment

Bold targets are useful, even if there is uncertainty around how they will be met.

I have followed developing countries' international commitments while working on climate policy with New Delhi think tanks and Tufts University, and I have participated in international negotiations as an observer.

India's mixed messaging on climate targets is part of a broader reluctance among large developing economies to accelerate their efforts to rein in climate change without finance from developed countries. Developed countries in 2015 promised to deliver $100 billion a year for decarbonization and adaptation in developing economies, but they have yet to meet that target.

Our research at Tufts University's Climate Policy Lab indicates that ambitious climate policies supported by international finance have a positive effect on GDP and employment. Communicating its national ambitions in its conditional NDC targets could have done more to attract the international finance India needs to fully implement them.

Viewpoint 5

> *"With plentiful hydropower and sunshine, Latin America already meets a quarter of its energy demand with renewables—nearly twice the global average."*

China Is Supporting the Rise of Renewable Energy in Other Countries, but Homegrown Innovation Is Also Important

Zdenka Myslikova and Nathaniel Dolton-Thornton

In this viewpoint, Zdenka Myslikova and Nathaniel Dolton-Thornton explain the role China has played in supporting the rise of renewable energy in Latin America, looking particularly at its impact on Chile. China has produced a large amount of wind and solar technologies that are used in Chile, and it has also invested in the critical minerals industry in the region. China is now a major player in the renewable energy supply chain, and though its influence on countries' energy systems has raised concerns, the authors argue that Latin American countries can use this to their benefit to create their own innovative energy strategies. Zdenka Myslikova is a postdoctoral scholar in clean energy innovation at Tufts University, where Nathaniel Dolton-Thornton is an assistant researcher in climate policy.

"'Global China' is a big part of Latin America's renewable energy boom, but homegrown industries and 'frugal innovation' are key," by Zdenka Myslikova and Nathaniel Dolton-Thornton, The Conversation, July 5, 2023, https://theconversation.com/global-china-is-a-big-part-of-latin-americas-renewable-energy-boom-but-homegrown-industries-and-frugal-innovation-are-key-208561.

As you read, consider the following questions:

1. What percent of the solar and wind technologies in Chile were produced by Chinese companies?
2. What mineral found in Chile is China interested in? What is Chile doing to protect this resource?
3. What is green hydrogen? What benefits can it offer?

The story of renewable energy's rapid rise in Latin America often focuses on Chinese influence, and for good reason. China's government, banks and companies have propelled the continent's energy transition, with about 90 percent of all wind and solar technologies installed there produced by Chinese companies. China's State Grid now controls over half of Chile's regulated energy distribution, enough to raise concerns in the Chilean government.

China has also become a major investor in Latin America's critical minerals sector, a treasure trove of lithium, nickel, cobalt, and rare earth elements that are crucial for developing electric vehicles, wind turbines, and defense technologies.

In 2018, the Chinese company Tianqi Lithium purchased a 23 percent share in one of Chile's largest lithium producers, Sociedad Química y Minera. More recently, in 2022, Ganfeng Lithium bought a major evaporative lithium project in Argentina for US$962 million. In April 2023, Brazilian President Luiz Inacio Lula da Silva and Chinese President Xi Jinping signed around 20 agreements to strengthen their countries' already close relationship, including in the areas of trade, climate change, and the energy transition.

China's growing influence over global clean energy supply chains and its leverage over countries' energy systems have raised international concerns. But the relationship between China and Latin America is also increasingly complicated as Latin American countries try to secure their resources and their own clean energy futures.

Alongside international investments, Latin American countries are fostering energy innovation cultures that are homegrown, dynamic, creative, often grassroots, and frequently overlooked. These range from sophisticated innovations with high-tech materials to a phenomenon known as "frugal innovation."

Chile Looks to the Future

Chile is an example of how Latin America is embracing renewable energy while trying to plan a more self-reliant future.

New geothermal, solar, and wind power projects—some built with Chinese backing, but not all—have pushed Chile far past its 2025 renewable energy goal. About one-third of the country is now powered by clean energy.

But the big prize, and a large part of China's interest, lies buried in Chile's Atacama Desert, home to the world's largest lithium reserves. Lithium, a silvery-white metal, is essential for producing lithium ion batteries that power most electric vehicles and utility-scale energy storage. Countries around the world have been scrambling to secure lithium sources, and the Chilean government is determined to keep control over its reserves, currently about one-half of the planet's known supply.

In April 2023, Chile's president announced a national lithium strategy to ensure that the state holds partial ownership of some future lithium developments. The move, which has yet to be approved, has drawn complaints that it could slow production.

However, the government aims to increase profits from lithium production while strengthening environmental safeguards and sharing more wealth with the country's citizens, including local communities impacted by lithium projects. Latin America has seen its resources sold out from under it before, and Chile doesn't intend to lose out on its natural value this time.

Learning from Foreign Investors

Developing its own renewable energy industry has been a priority in Chile for well over a decade, but it's been a rough road at times.

In 2009, the government began establishing national and international centers of excellence—Centros de Excelencia Internacional—for research in strategic fields such as solar energy, geothermal energy and climate resilience. It invited and cofinanced foreign research institutes, such as Europe's influential Fraunhofer institute and France's ENGIELab, to establish branches in Chile and conduct applied research. The latest is a center for the production of lithium using solar energy.

The government expected that the centers would work with local businesses and research centers, transferring knowledge to feed a local innovation ecosystem. However, reality hasn't yet matched the expectations. The foreign institutions brought their own trained personnel. And except for the recently established institute for lithium, officials tell us that low financing has been a major problem.

Chile's Startup Incubator and Frugal Innovation

While big projects get the headlines, more is going on under the radar.

Chile is home to one of the largest public incubators and seed accelerators in Latin America, StartUp Chile. It has helped several local startups that offer important innovations in food, energy, social media, biotech, and other sectors.

Often in South America, this kind of innovation is born and developed in a resource-scarce context and under technological, financial and material constraints. This "frugal innovation" emphasizes sustainability with substantially lower costs.

For example, the independent Chilean startup Reborn Electric Motors has developed a business converting old diesel bus fleets into fully electric buses. Reborn was founded in 2016 when the national electromobility market in Chile was in its early stages, before China's BYD ramped up electric bus use in local cities.

Reborn's retrofitted buses are both technologically advanced and significantly cheaper than their Chinese counterparts. While BYD's new electric bus costs roughly US$320,000, a retrofitted equivalent from Reborn costs roughly half, around $170,000.

The company has also secured funding to develop a prototype for running mining vehicles on green hydrogen.

Bolivia's "tiny supercheap EV" developed by homegrown startup Industrias Quantum Motors is another example of frugal innovation in the electric vehicles space. The startup aspires to bring electric mobility widely to the Latin American population. It offers the tiniest EV car possible, one that can be plugged into a standard wall socket. The car costs around $6,000 and has a range of approximately 34 miles (55 kilometers) per charge.

Phineal is another promising Chilean company that offers clean energy solutions, focusing on solar energy projects. Its projects include solar systems installation, electromobility technology and technology using blockchain to improve renewable energy management in Latin America. Many of these are highly sophisticated and technologically advanced projects that have found markets overseas, including in Germany.

Looking Ahead to Green Hydrogen

Chile is also diving into another cutting-edge area of clean energy. Using its abundant solar and wind power to produce green hydrogen for export as a fossil fuel replacement has become a government priority.

The government is developing a public-private partnership of an unprecedented scale in Chile for hydrogen production and has committed to cover 30 percent of an expected $193 million public and private investment, funded in part by its lithium and copper production. Some questions surround the partnership, including Chile's lack of experience administering such a large project and concerns about the environmental impact. The government claims Chile's green energy production could eventually rival its mining industry.

With plentiful hydropower and sunshine, Latin America already meets a quarter of its energy demand with renewables – nearly twice the global average. Chile and its neighbors envision those numbers only rising.

Viewpoint 6

> *"Industrialized countries have been reluctant to commit funding, concerned it could lead to being legally liable for the impacts of climate change. At this year's climate talks, developing countries say it's a crucial part of climate justice."*

The Loss and Damage Caused by Climate Change Makes It Too Expensive to Continue Using Fossil Fuels

Lauren Sommer

As the devastation caused by climate change and the extreme weather events that are a result of it become increasingly common, some developing countries argue that the developed countries responsible for most greenhouse gas emissions should help cover the cost of the damage done by climate change. In this viewpoint, Lauren Sommer discusses the arguments being made. Developed countries such as the U.S. are reluctant to be held financially responsible for the damage done to other countries, as they fear this would mean admitting liability for the issue, which could be very expensive in the long run. However, it is clear that the impacts of climate change have caused a massive amount of loss and damage that developing countries are not financially equipped to address, making the continued use of fossil fuels a costly—and deadly—choice. Lauren Sommer is a correspondent for NPR's climate desk.

As you read, consider the following questions:

1. What countries that have suffered devastating weather events due to climate change are mentioned in this viewpoint?
2. What recent scientific advancements have helped developing countries make a stronger case for a loss and damage fund, according to this viewpoint?
3. According to Maarten van Aalst, why is it difficult to assign blame for damage caused by extreme weather events?

Barbados Prime Minister Mia Mottley wants richer countries to stop throwing garbage in her yard and then telling her to clean it up.

The garbage, in this case, is greenhouse gas emissions that fuel more extreme storms and hurricanes, causing widespread destruction which can cost billions of dollars. At the Glasgow climate negotiations, Mottley is leading a push for richer countries to compensate poorer ones for the "loss and damage" caused by climate change.

Their argument is this: developed countries, like the U.S and those in the European Union, are responsible for most of the heat-trapping emissions pumped into the atmosphere since the Industrial Revolution. Developing countries have lower emissions, but are still bearing the brunt of a hotter climate through more severe heat waves, floods, and droughts.

"It is unjust and it is immoral," Mottley said at the summit. "It is wrong."

To help compensate for that, developing countries are asking richer ones to contribute to a loss and damage fund. The money could offer payment for things that are irrevocably lost, like lives or the extinction of species. It could also help countries with the cost of rebuilding after storms, replacing damaged crops, or relocating entire communities at risk.

While loss and damage was discussed at the Paris climate talks in 2015, progress has been slow. Industrialized countries have been reluctant to commit funding, concerned it could lead to being legally liable for the impacts of climate change. At this year's climate talks, developing countries say it's a crucial part of climate justice.

"Providing finance for loss and damage is the very least that wealthy countries can and should do," says Raeed Ali, a climate activist from Fiji and part of the Loss and Damage Youth Coalition. "But to do this, they will have to acknowledge that they are responsible for this. And I think that is something they are not willing to do."

Few Resources to Recover from Disasters

For countries where much of the population lives in poverty, extreme weather can be a devastating blow. Individuals have little savings to rebuild, while governments with few resources struggle to secure the millions of dollars needed to help communities recover.

"In Fiji, we are at the forefront of the climate crisis," Ali says. "So every single person knows about climate change because it's a daily reality for us."

Ali says while his grandparents only remember experiencing one category 5 cyclone in their lives, he's already seen three. And with rising sea levels threatening to make villages uninhabitable, a handful in Fiji have already been relocated, and more than 40 are slated to be moved, he said.

Some countries are being hit by back-to-back disasters. In 2015, the Caribbean island nation of Dominica was hit by tropical storm Erika, causing more than $400 million in damage, equal to 90 percent of the country's gross domestic product. Two years later, Hurricane Maria slammed the island, damaging 90 percent of the country's housing stock.

In the Gambia in West Africa, where the majority of people in rural areas depend on agriculture, crop failures can be catastrophic.

Because the country's main river flows into the ocean, rising sea levels are pushing saltwater farther and farther upriver, making it harder to get freshwater.

"It's really affecting the farming communities," says Isatou Camara, a climate negotiator for the Gambia in West Africa. "Because of sea level rise, we have saltwater intruding into our river which is affecting farm production, especially rice farming which is usually done along the riverside."

Developed countries have promised $100 billion per year in "climate finance" to help poorer nations reduce their emissions through things like renewable energy and sustainable agriculture. They've yet to fully deliver on that goal, since each country determines its own contribution.

But many developing countries say that funding doesn't help with the climate impacts they're already experiencing, which is why a separate loss and damage fund is needed. In 2020, natural disasters caused $210 billion in damage worldwide.

Assigning Liability for Climate Change

At the 2015 Paris climate summit, countries signed an agreement recognizing the need to address loss and damage. But developed countries pushed to include language that specified it did not "provide a basis for any liability." They feared that admitting responsibility for their share of heat-trapping pollution would expose them to paying developing nations every time a disaster hit.

"It's always something developed countries have been very cautious about exactly because they don't want it to be a precedent for international courts," said Maria Antonia Tigre, a fellow at Columbia University's Sabin Center for Climate Change Law. "They really do want to avoid that responsibility because it can be endless."

Recent advances in climate change science have upped the pressure. While many climate studies examine long-term trends, researchers at the World Weather Attribution initiative study whether climate change has amplified an extreme weather event in the weeks or months after it hits.

Hurricane Harvey, which released a deluge of rain on Houston in 2017, was made 15 percent more intense by climate change, they found. This past summer, the severe heat wave in the Pacific Northwest that caused dozens of deaths was virtually impossible without the added boost of human-caused greenhouse gas emissions.

"The fact that we are able to pinpoint the climate fingerprint in specific things that hurt us today, I think, is a very important element in the current loss and damage conversation," says Maarten van Aalst, director of the Red Cross Red Crescent Climate Centre and a scientist with the World Weather Attribution initiative. "It has changed the conversation over the past years."

Van Aalst is quick to point out that assigning blame is a complex question. While a storm's destruction is caused by its strength, it also occurs if the buildings and homes aren't built to handle storms. Some communities aren't designed to endure the kinds of events that occurred even before the effects of climate change started to be felt.

Still, court cases are underway around the world to establish liability for climate change, either seeking damages from governments or fossil fuel companies.

"The uncertainty of science was one of the main arguments that was always used by states to avoid responsibility," says Tigre. "And that's a bridge that's now crossed."

No Show of Funding for Loss and Damage

At the Glasgow summit, Scotland announced a major milestone for addressing loss and damage needs. It offered 2 million pounds in funding, the first of its kind.

"That is in the right direction," said Sonam Wangdi of Bhutan, who chairs a group of the 46 poorer countries at the talks. "It's going to be very clear that there should be separate funds for loss and damage, and it should not be mixed with all the other funds."

Still, other funding commitments haven't followed yet. While the U.S. has formally recognized the need to address loss and damage, a senior U.S. official says the country doesn't support creating a dedicated new fund.

Instead of waiting for voluntary offers, Barbados Prime Minister Mottley is proposing using 1 percent tax on sales revenues from fossil fuels, which she estimates could raise $70 billion per year. Some are simply hoping that countries invest in the United Nation's Santiago Network, which was set up in 2019 to handle loss and damage issues. But without enough staffing and funding, it exists mainly symbolically today.

Even if developed countries offer new support for a loss and damage fund, they could still be held liable for the impacts of climate change. At a press conference in Glasgow, the island nations of Tuvalu and Antigua and Barbuda announced they're forming a new commission to enable small island countries to seek compensation through international courts.

Periodical and Internet Sources Bibliography

The following articles have been selected to supplement the diverse views presented in this chapter.

"Can the World Run on Renewable Energy?" Knowledge at Wharton, April 23, 2015. https://knowledge.wharton.upenn.edu/article/can-the-world-run-on-renewable-energy/.

Sibi Arasu, "India's Energy Future Is Looking Green, Report Says," Associated Press, October 14, 2022. https://apnews.com/article/business-climate-and-environment-government-politics-60b7c65cca2c38c26a960d14732bb8bb.

Adrienne Bernhard, "How Limitless Green Energy Would Change the World," BBC, October 6, 2022. https://www.bbc.com/future/article/20221006-what-would-happen-if-we-had-limitless-green-energy.

Fatih Birol and Amitabh Kant, "India's Clean Energy Transition Is Rapidly Underway, Benefiting the Entire World," *Times of India*, January 9, 2022. https://timesofindia.indiatimes.com/blogs/toi-edit-page/on-track-for-2070-net-zero-target-indias-clean-energy-transition-is-rapidly-underway-benefiting-the-entire-world/.

Harrison Dreves, "At a Glance: How Renewable Energy Is Transforming the Global Electricity Supply," National Renewable Energy Laboratory, September 26, 2023. https://www.nrel.gov/news/program/2023/how-renewable-energy-is-transforming-the-global-electricity-supply.html.

Eva Hamrud, "Is 100% Renewable Energy Feasible For Entire Countries? Why, Yes Actually," *ScienceAlert*, July 10, 2021. https://www.sciencealert.com/these-climate-experts-say-100-renewable-energy-is-completely-feasible-for-entire-countries.

Yusuf Khan, "Net Zero Is Still Possible, but Clean Energy Spending Must Go Faster," *Wall Street Journal*, September 26, 2023. https://www.wsj.com/articles/net-zero-is-still-possible-but-clean-energy-spending-must-go-faster-a14fca86.

Lauri Myllyvirta, "Clean Energy Was Top Driver of China's Economic Growth in 2023," *Carbon Brief*, January 25, 2024. https://www.

carbonbrief.org/analysis-clean-energy-was-top-driver-of-chinas-economic-growth-in-2023/.

Christina Nunez, "Renewable Energy, Explained," *National Geographic*, January 30, 2019. https://www.nationalgeographic.com/environment/article/renewable-energy?loggedin=true&rnd=1708719071059.

Lu Shuo, "Is China Really Leading the Clean Energy Revolution? Not Exactly," the *Guardian*, July 6, 2023. https://www.theguardian.com/commentisfree/2023/jul/06/china-clean-energy-revolution-coal-power.

OPPOSING VIEWPOINTS® SERIES

Chapter 2

Can Renewable and Alternative Energy Solve the Climate Crisis?

Chapter Preface

Climate change is a massive problem. Most efforts to address it have involved turning to renewable sources of energy, such as wind and solar power. However, as with most big problems, the solution is not simple. Implementing renewable energy sources when fossil fuels are already embedded in the world's economies and infrastructure makes it difficult to change. And aside from the economics, the environmental consequences of renewable energy can be complicated as well. On top of that, will turning to renewables be enough? Can alternative and renewable energy *really* solve the climate crisis? That's the question the authors in this chapter try to answer.

The chapter opens with a viewpoint that argues that renewable sources of energy are indeed the answer to the climate crisis. What we must do, these authors write, is accelerate the movement toward renewables. However, the author of the second viewpoint is not so sure. He writes that while renewables are important, even essential, they are not enough. We must begin removing carbon from the atmosphere, while at the same time changing the economic structure that is dependent on the fossil fuel model.

Then we hear from a couple of authors who argue that in order to prevent climate disaster, we must completely restructure our relationship with the rest of the natural world and carefully consider the choices we make. The final viewpoint argues that while transitioning away from fossil fuels will be a big change, it is not a complex one, and it's clear that it's the only way to address climate change.

Viewpoint 1

> *"The IPCC reports make clear that the future inevitably involves more and larger climate-related transformations. The question is what the mix of good and bad will be in those transformations."*

Climate Change Mitigation Is Essential, and Renewable Energy Plays an Essential Role in It

Robert Lempert and Elisabeth Gilmore

In this viewpoint, Robert Lempert and Elisabeth Gilmore discuss the 2023 reports from the Intergovernmental Panel on Climate Change (IPCC), which is a body of the UN that works to advance knowledge on climate change and its causes. The report indicated that changes in response to climate change will be inevitable for society, but also provided potential solutions to mitigate and adjust to climate change. One of the main ways climate change can be mitigated is by switching to renewable energy sources, which will reduce emissions. So far there hasn't been enough of a shift to renewable energy, and the authors speculate that this is because of pushback from fossil fuel industries as well as fear of change. However, in order to meet the changes impacting the planet, people around the world must make adjustments now. Robert Lempert is a principal researcher at RAND and a professor of policy analysis at Pardee RAND Graduate

School. Elisabeth Gilmore is an associate professor of climate change, technology, and policy at Carleton University.

As you read, consider the following questions:

1. According to this viewpoint, what were the positive impacts of the Industrial Revolution? What were the negative effects?
2. What negative changes do the authors anticipate if greenhouse gas emissions continue at a high rate?
3. According to this viewpoint, what suggestions does the IPCC have for making positive transformations?

It's easy to feel pessimistic when scientists around the world are warning that climate change has advanced so far, it's now inevitable that societies will either transform themselves or be transformed. But as two of the authors of a recent international climate report, we also see reason for optimism.

The latest reports from the Intergovernmental Panel on Climate Change, including the synthesis report released March 20, 2023, discuss changes ahead, but they also describe how existing solutions can reduce greenhouse gas emissions and help people adjust to impacts of climate change that can't be avoided.

The problem is that these solutions aren't being deployed fast enough. In addition to pushback from industries, people's fear of change has helped maintain the status quo.

To slow climate change and adapt to the damage already underway, the world will have to shift how it generates and uses energy, transports people and goods, designs buildings, and grows food. That starts with embracing innovation and change.

Fear of Change Can Lead to Worsening Change

From the Industrial Revolution to the rise of social media, societies have undergone fundamental changes in how people live and understand their place in the world.

Some transformations are widely regarded as bad, including many of those connected to climate change. For example, about half the world's coral reef ecosystems have died because of increasing heat and acidity in the oceans. Island nations like Kiribati and coastal communities, including in Louisiana and Alaska, are losing land into rising seas.

Other transformations have had both good and bad effects. The Industrial Revolution vastly raised standards of living for many people, but it spawned inequality, social disruption, and environmental destruction.

People often resist transformation because their fear of losing what they have is more powerful than knowing they might gain something better. Wanting to retain things as they are—known as status quo bias—explains all sorts of individual decisions, from sticking with incumbent politicians to not enrolling in retirement or health plans even when the alternatives may be rationally better.

This effect may be even more pronounced for larger changes. In the past, delaying inevitable change has led to transformations that are unnecessarily harsh, such as the collapse of some 13th-century civilizations in what is now the U.S. Southwest. As more people experience the harms of climate change firsthand, they may begin to realize that transformation is inevitable and embrace new solutions.

A Mix of Good and Bad

The IPCC reports make clear that the future inevitably involves more and larger climate-related transformations. The question is what the mix of good and bad will be in those transformations.

If countries allow greenhouse gas emissions to continue at a high rate and communities adapt only incrementally to the resulting climate change, the transformations will be mostly forced and mostly bad.

For example, a riverside town might raise its levees as spring flooding worsens. At some point, as the scale of flooding increases, such adaptation hits its limits. The levees necessary to hold back the water may become too expensive or so intrusive that they undermine any benefit of living near the river. The community may wither away.

The riverside community could also take a more deliberate and anticipatory approach to transformation. It might shift to higher ground, turn its riverfront into parkland while developing affordable housing for people who are displaced by the project, and collaborate with upstream communities to expand landscapes that capture floodwaters. Simultaneously, the community can shift to renewable energy and electrified transportation to help slow global warming.

Optimism Resides in Deliberate Action

The IPCC reports include numerous examples that can help steer such positive transformation.

For example, renewable energy is now generally less expensive than fossil fuels, so a shift to clean energy can often save money. Communities can also be redesigned to better survive natural hazards through steps such as maintaining natural wildfire breaks and building homes to be less susceptible to burning.

Land use and the design of infrastructure, such as roads and bridges, can be based on forward-looking climate information. Insurance pricing and corporate climate risk disclosures can help the public recognize hazards in the products they buy and companies they support as investors.

No one group can enact these changes alone. Everyone must be involved, including governments that can mandate and incentivize changes, businesses that often control decisions about greenhouse gas emissions, and citizens who can turn up the pressure on both.

Transformation Is Inevitable

Efforts to both adapt to and mitigate climate change have advanced substantially in the last five years, but not fast enough to prevent the transformations already underway.

Doing more to disrupt the status quo with proven solutions can help smooth these transformations and create a better future in the process.

Viewpoint 2

> *"Yes, massive solar and wind parks are being built around the world, but they haven't yet changed the business models of Shell, BP, and other fossil fuel giants."*

We Can't Count on Renewable Energy to Save Us

Steffen Böhm

Unlike the authors of the previous viewpoint, Steffan Böhm is not so optimistic about renewable energy. Though he says moving to renewable sources of energy is important, it is not enough. He explains that the issue of land shortage must be addressed before solar and wind energy can become dominant sources of energy. This shift would also require a large amount of metals like aluminum, steel, and copper. These issues present a number of challenges, and furthermore, Böhm argues that in addition to preventing more carbon from being emitted, efforts must be made to remove carbon that already exists in the atmosphere. Steffan Böhm is a professor of organization and sustainability at the University of Exeter Business School in the UK.

As you read, consider the following questions:

1. According to this viewpoint, what is the problem with the Paris Climate Agreement?
2. What are the "hard facts" about renewables that Böhm says we must face?
3. What solution does Böhm suggest?

The Paris climate agreement has now officially come into force. Although Donald Trump and other climate change deniers have vowed to abandon it, most have hailed the agreement as a huge success and a significant milestone in our quest to limit the effects of global climate change.

But here's the problem: many climate experts warn that the commitments made at Paris still fall far short of what is required to halt global warming at the 2°C mark, never mind reversing the growth of greenhouse gases in the atmosphere. The simple truth is that the Paris agreement is blind to the fundamental, structural problems that prevent us from decarbonising our economies to the radical extent needed.

Take renewable energy. Among the most progressive leaders in business, government, and NGOs there is a shared belief that, if only we could switch off the fossil fuel tap and quickly transition towards renewable energy sources, we still have a chance to save the world from runaway climate change. All that's needed is massive investment in wind, solar, geothermal, and other renewables. International agreements such as those reached in Paris are what makes those investments possible, providing business confidence and policy commitment. While I feel part of this group of progressives, there are some hard facts that cannot be ignored.

Fossil Fuel Still Dominates

First, the renewable schemes to date have largely been at the expense of unpopular nuclear installations, while the global share of fossil fuel-generated energy consumption remains at about 80 to

85 percent: just where it's been since the early 1970s. Yes, massive solar and wind parks are being built around the world, but they haven't yet changed the business models of Shell, BP, and other fossil fuel giants. On the contrary, they feel more secure than ever to invest in fossil fuel sources, particularly gas, which they see as a "transition fuel"—here to stay until at least 2050, they say.

Land Shortage

Second, the massive amounts of land required for installing gigawatts of solar and wind power will destroy natural habitats and take away valuable farmland. This is already evident in the way existing biomass production schemes—forests in the United States for instance, sugar cane in Brazil, or palm oil in Malaysia—have had serious environmental and social side effects to the extent that they have been labelled as "greenwash."

There simply isn't enough accessible land for all the solar or wind farms that would be needed to transition to a renewable future. Wherever renewables have been developed at the "mega" level, they end up bulldozing, quite literally, people and wildlife. And generally it's the poorest, usually rural, communities who are disproportionately affected, given that their land values are lowest and existing users have little power or formal land rights. For example, large- scale hydroelectric dam projects, currently the greatest source of renewable energy, have destroyed many human communities and flooded irreplaceable natural habitats.

Yes, offshore wind can fill some of the gaps, but it is more expensive to build and maintain than onshore, and the generated energy has to be transmitted over long distances.

Heavy on Metals

Third, as French scientist Olivier Vidal and his colleagues recently pointed out, the shift to renewable energy will "replace one nonrenewable resource (fossil fuels) with another (metals and minerals)." Vidal estimates that 3,200 million tonnes of steel, 310 million tonnes of aluminum and 40 million tonnes

Nuclear Power May be Needed to Hasten Decarbonization

With the adoption of the Paris Agreement in 2015, almost all Parties to the United Nations Framework Convention on Climate Change (UNFCCC) agreed to prepare nationally determined contributions (NDCs) to control GHG emissions and limit the increase of global mean surface temperature by the end of the century to below 2°C relative to pre-industrial levels. Since then, increasing scientific understanding of the significant risks associated with warming of 2°C, along with increasing societal concern, have established the need for more urgent and ambitious action to avoid the worst impacts of climate change, by limiting warming to 1.5°C.

To reach this goal, carbon dioxide (CO2) emissions from electricity generation must fall to nearly zero by the middle of this century, even as electricity needs worldwide continue to grow and expand in end-uses such as transportation, heating and industrial energy use.

Nuclear power is a low-carbon source of energy. In 2018, nuclear power produced about 10 percent of the world's electricity. Together with the expanding renewable energy sources and fuel switching from coal to gas, higher nuclear power production contributed to the levelling of global CO2emissions at 33 gigatonnes in 2019. Clearly, nuclear power – as a dispatchable low carbon source of electricity – can play a key role in the transition to a clean energy future.

As part of the capacity building process for energy system analysis and planning, the IAEA provides assistance to Member States for the evaluation of the role of nuclear energy in national climate change mitigation strategies through the Technical Cooperation programme and Coordinated Research Projects. For this purpose, a comprehensive set of IAEA tools and methodologies are available to Member States.

"Nuclear Power and Climate Change: Decarbonization," International Atomic Energy Agency

of copper would be needed to build the latest generations of wind and solar facilities. Together with demand from electric vehicle manufacturers, a worldwide renewables boom would rely on a 5 percent to 18 percent annual increase in global production of minerals for the next 40 years.

Similarly startling projections are made for other materials oiling the wheels of green capitalism, including silver, lithium, copper, silicon, gallium, and the rare earths. In many cases, supplies of these raw materials are already dwindling. The Toyota Prius, for example, one of the greenest cars on the market, relies on a range of very dirty rare earth minerals, the excavation and processing of which has devastated large areas of Inner Mongolia in China.

Removing Carbon

Lastly, the climate challenge is so urgent and huge that we actually need to remove carbon from the atmosphere, rather than just switching to renewables. That's the view of prominent climate scientist James Hansen, the former head of NASA's Goddard Institute for Space Studies, who has shown that, even if we switched to zero-carbon energy sources today, we would still be facing a serious climate challenge for centuries to come.

What this all means is that the Paris agreement doesn't go far enough. In fact, it might give us the impression of moving in the right direction, but actually the pledged actions are so far off what is needed, it spreads false hope.

So, what is needed then?

- A realization that simply switching to renewables alone will not solve the climate change problem.
- We need to start removing carbon from the atmosphere.
- We need to tackle the demand side. We cannot simply assume that relentless economic growth is compatible with a green future.

These points raise uncomfortable questions that only those who can think and act against the grain dare to ask. I'm not saying that we shouldn't transition to renewable energy. Not at all. But that alone will not save the climate. The world's climate experts and leaders in business, government, and NGOs, who are about to gather in Marrakesh for yet another UN conference, would do well in starting to engage with this uncomfortable truth.

VIEWPOINT 3

> *"It's impossible to take effective action against global warming without addressing our impact on the rest of the living world, and vice versa."*

Preventing a Climate Disaster Will Take More than Decarbonizing

Christian de Perthuis and Édouard Civel

In this viewpoint, the authors argue that in order to prevent the worst effects of climate change, we need to go beyond reducing carbon emissions. We need to completely change how we interact with the rest of the natural world. Forests and oceans play important roles in absorbing CO_2—they're even referred to as carbon sinks—so by protecting these ecosystems we can also help reduce climate change. Christian de Perthuis is a professor of economics and chair of climate economics at the Paris-Dauphine University. Édouard Civel is a researcher at the Square Research Center at Paris-Dauphine University.

As you read, consider the following questions:

1. What is the Intergovernmental Sciences Policy Platform on Biodiversity and Ecosystem Services?

"Why we won't be able to prevent climate breakdown without changing our relationship to the rest of the living world", by Christian de Perthuis and Édouard Civel, The Conversation, September 5, 2023. https://theconversation.com/why-we-wont-be-able-to-prevent-climate-breakdown- without-changing-our-relationship-to-the-rest-of-the-living-world-212401.

2. What is meant by "living carbon"?
3. According to this viewpoint, how will increasing biodiversity help with climate change?

Because of its growing impact on society, global warming has taken centre stage in the public debate. While most of us have not read the reports by the Intergovernmental Panel on Climate Change (IPCC), heat waves, intensifying storms and the multiplication of extreme events remind us of the scale of climate disruption and the urgency of action.

Despite being documented by the Intergovernmental Sciences Policy Platform on Biodiversity and Ecosystem Services (IPBES), the equivalent of the IPCC for biodiversity, we know little about how biodiversity erosion might affect us and the rest of the planet. Its links and interactions with climate change are underestimated, and any policy to address either in isolation will miss the mark. It's impossible to take effective action against global warming without addressing our impact on the rest of the living world, and vice versa.

Fossil Carbon, Living Carbon

IPCC scientists have been explaining since their first assessment report (1990) that climate change is a stock problem. To halt global warming, it is not enough to slash greenhouse gas emissions. We need to stabilise their stock in the atmosphere. To achieve reach net zero we must reduce emissions—the inflow into the stock—to the level of the outflow, which is made up of CO_2 absorption by carbon sinks (forests and oceans) and the elimination of non-CO_2 greenhouse gases at the end of their life cycle.

This requires that we adopt a two-pronged plan, aimed both at cutting down our reliance on both fossil and living carbon. The former feeds the vast majority of the world's pollution, with coal, oil, and natural gas accounting for 70 percent of the world's greenhouse gas emissions.

Tackling it will require that we take on the so-called energy transition.

On the other hand, a quarter of greenhouse gas emissions come from "living carbon", mainly as a result of specific agricultural emissions (unrelated to fossil fuel use) and tropical deforestation and other land use changes that erode carbon sinks. There is no way to achieve carbon neutrality without a profound transformation in the use of living resources, to ensure the reflux of agricultural emissions and better protection of carbon sinks. This is the challenge of what we might call the agroclimatic transition.

One of the major difficulties of the ecological transition is to carry out these two transformations simultaneously, as they involve distinct economic mechanisms. For fossil carbon, we need to introduce scarcity by reducing the use of coal, oil, and natural gas to the absolute minimum. For living carbon, we need to reinvest in the diversity of ecosystems to reduce agricultural emissions and protect carbon sinks as part of a bioeconomy.

From Adding to Subtracting

Since the start of the Industrial Revolution, energy transitions have followed one another. They have all involved adding new energy sources to a system initially based on the use of biomass. The result has been a massive increase in the amount of energy used worldwide.

The climate is forcing us to break with this logic. Lowering emissions is not a matter of adding decarbonised sources to the energy system. It's about removing fossil fuels. We need to switch from a logic of addition to one of subtraction.

From an economic viewpoint, this means massively reconverting brown assets linked to the production or use of fossil fuels, through a double movement of investment in green and disinvestment in brown. The heaviest cost for the economic system is not the hundreds of billions invested in wind or solar farms, battery gigafactories or hydrogen electrolysers. It's the cost of disinvestment that forces us to downgrade or reconvert brown assets: financial assets, of course, but also physical assets and, above all, the human assets on which the energy transition depends.

Multiple instruments will have to be called upon to bring about such a transformation.

Pricing carbon from fossil fuel use is a key way to reflect the increasing scarcity of the atmospheric capacity to store carbon. Whether obtained through taxation or emission trading schemes, such taxation raises the cost of using fossil fuels, without returning the resulting rents to producers, as happens, for example, when oil prices soar on energy markets. On the demand side, it is a powerful stimulus to energy efficiency and sufficiency; on the supply side, it encourages a shift away from carbon assets.

The main difficulty with fossil carbon taxation lies in controlling its distributive impact. As the "gilets jaunes" protests in France showed, fossil carbon taxation without redistribution to the most vulnerable poses more problems than it solves. Only a redistributive carbon tax will be socially acceptable. Similarly, if carbon pricing is to be extended on an international scale, the proceeds must be returned on a massive scale to the countries of the South.

The distributional impacts of regulated carbon markets should also not be underestimated. Within the European Union, the extension of the emission trading scheme to the transport and buildings sector will increase household energy bills. This is why the proceeds from allowances sales at auction must be redistributed to the most vulnerable households via a "social fund" which will be the pillar of the regulation to be put in place.

While fossil carbon taxation accelerates the energy transition, negative carbon taxes—in other words, fossil fuel subsidies – delay it. Following the outbreak of war in Ukraine, these subsidies reached unprecedented levels in the European Union, with the multiplication of "tariff shields" erected as a matter of urgency to protect Europeans from the worst of the cost of living crisis.

Another pernicious form of subsidy to fossil fuels is the free allocation of CO_2 allowances in the European trading scheme, which hampers the emergence of a green industry, a lever for the competitiveness of tomorrow's Europe.

Investing in the Diversity of Living Beings

Let's imagine for a moment that the world has eradicated all use of fossil fuels in 2050. Would we automatically be in a situation of climate neutrality? Everything depends on what has been achieved on the second front of the transition, that of living carbon, the source of a quarter of the world's greenhouse gas emissions.

Pricing fossil carbon is hardly useful for the agroclimatic transition. Worse, it could even prove counterproductive: using a CO_2 price based on energy criteria, it would become profitable to transform the Amazon rainforest (or the centuries-old oaks of the French Tronçay forest) into short rotation coppice to produce energy! The reason is simple. Agroclimatic transformation means finding ways to reinvest in biological diversity, in other words, in the abundance of living things. But the price of CO_2 does not reflect the value of this diversity. We therefore need to use other instruments, which are more complex to implement.

On land, forests are the main carbon sink. Their capacity to soak up atmospheric CO_2 is weakened by a combination of climatic and anthropogenic factors. In France, for example, the CO_2 storage capacity of forests has been divided by three since 2005, mainly due to climatic factors. There is therefore an urgent need to adapt forest management methods in anticipation of the severity of tomorrow's climates. Worldwide, the main anthropogenic impact on forests is tropical deforestation. Its main cause is the expansion of land for crops and livestock. This is why the key to halting deforestation lies in changing agricultural practices.

The Key Issues of Agriculture and Food

The impact of farming systems on the net balance of greenhouse gas emissions is not limited to deforestation. Depending on the techniques used, farming systems may themselves release carbon into the atmosphere (deep ploughing, draining of wet soils, etc.) or, on the contrary, store it in living soils (conservation agriculture, agroforestry, etc.). The former erode biodiversity by specialising farmers according to industrial-type logics. The latter use living diversity to intensify production and regenerate the natural environment.

These agroecological techniques also make it possible to better withstand tougher climatic conditions, while reducing methane and nitrous oxide emissions from agricultural sources. In economic terms, their promotion requires investment in innovation, research, and development, the establishment of dedicated farm advisory networks, and, above all, incentivisation to reward farmers for the ecosystem services they provide to society. This is not something that happens spontaneously on the market. It requires public intervention and dedicated funding.

As in the case of energy, the agroclimatic transition implies, on the demand side, that we consume smarter and less. The foods we eat have contrasting climate footprints. There can be no successful agroclimatic transition without finding ways to dramatically reduce emissions associated with the most polluting ingredients, including industrially processed foods and animal products, especially those from ruminant breeding. The use of food rations might be one way of achieving this, according to the recommendations of the world's health authorities.

Remembering the Ocean

Last but not least, the agroclimatic transition will have to take into account the management of the oceans and marine biodiversity, which are currently the blind spots of climate policies.

Global warming and certain human practices (overfishing, pollutant runoff, etc.) are altering marine biodiversity, a crucial component in the storage of CO_2 by the oceans. Protecting the ocean sink is vital to stabilise tomorrow's climate: it is estimated that the continental biosphere contains four times more carbon than the atmosphere. For the oceans, it's 47 times.

Viewpoint 4

> *"We, as individuals, must change our consumption habits and pressure those who represent us—our employers, our politicians—to move rapidly to a low-carbon world."*

Consumers Can Play an Important Part in Decarbonization

United Nations Environment Programme

In this viewpoint from the United Nations Environment Programme (UNEP), the author explains ways in which individuals can make personal decisions that can help reduce greenhouse gas emissions. This includes speaking out against carbon pollution and lobbying politicians and businesses to make more environmentally friendly choices. It can also involve choosing to use transportation powered by renewable energy and possibly even switching to a renewable energy provider at home. The United Nations Environment Programme (UNEP) is responsible for coordinating responses to environmental issues within the United Nations system.

As you read, consider the following questions:

1. According to the environmental platform Count Us In, how much could global carbon emissions be reduced if

"10 ways you can help fight the climate crisis," United Nations Environment Programme (UNEP), May 4, 2022. Reprinted by permission.

1 billion people decided to make changes to reduce their carbon footprints?
2. According to this viewpoint, what are some sources of transportation that create fewer greenhouse gas emissions?
3. What percent of global carbon emissions are from the fashion industry?

The evidence is irrefutable: unless we act immediately to reduce greenhouse gas emissions, we will not be able to stave off the worst consequences of climate change.

The world is already 1.2°C warmer than pre-industrial times and every fraction of a degree counts. Research shows that with 2°C of global warming we will have more intense droughts and more devastating floods, more wildfires, and more storms.

As United Nations Secretary-General António Guterres said at the recent UN Climate Change Conference of the Parties (COP26), "Our fragile planet is hanging by a thread. We are still knocking on the door of climate catastrophe. It is time to go into emergency mode—or our chance of reaching net-zero will itself be zero."

The outlook can seem depressing. But the good news is that there is a lot we can still do as individuals to change this narrative.

"The climate emergency demands action from all of us. We need to get to net zero greenhouse gas emissions by 2050 and everyone has a role to play," said Niklas Hagelberg, UNEP's climate change coordinator. "We, as individuals, must change our consumption habits and pressure those who represent us—our employers, our politicians—to move rapidly to a low-carbon world."

Here are 10 ways you can be part of the climate solution:

Spread the Word

Encourage your friends, family, and coworkers to reduce their carbon pollution. Join a global movement like Count Us In, which aims to inspire 1 billion people to take practical steps and challenge their leaders to act more boldly on climate. Organizers of the platform say that if 1 billion people took action, they could reduce as much as 20 percent of global carbon emissions. Or you could sign up to the UN's #ActNow campaign on climate change and sustainability and add your voice to this critical global debate.

Keep Up the Political Pressure

Lobby local politicians and businesses to support efforts to cut emissions and reduce carbon pollution. #ActNow Speak Up has sections on political pressure and corporate action—and Count Us In also has some handy tips for how to do this. Pick an environmental issue you care about, decide on a specific request for change and then try to arrange a meeting with your local representative. It might seem intimidating but your voice deserves to be heard. If humanity is to succeed in tackling the climate emergency, politicians must be part of the solution. It's up to all of us to keep up with the pressure.

Transform Your Transport

Transport accounts for around a quarter of all greenhouse gas emissions and across the world, and many governments are implementing policies to decarbonize travel. You can get a head start: leave your car at home and walk or cycle whenever possible. If the distances are too great choose public transport, preferably electric options. If you must drive, offer to carpool with others so that fewer cars are on the road. Get ahead of the curve and buy an electric car. Reduce the number of long-haul flights you take.

Rein in Your Power Use

If you can, switch to a zero-carbon or renewable energy provider. Install solar panels on your roof. Be more efficient: turn your heating down a degree or two, if possible. Switch off appliances and lights when you are not using them and better yet buy the most efficient products in the first place. (Hint: this will save you money!). Insulate your loft or roof: you'll be warmer in the winter, cooler in the summer and save some money too.

Tweak Your Diet

Eat more plant-based meals – your body and the planet will thank you. Today, around 60 percent of the world's agricultural land is used for livestock grazing and people in many countries are consuming more animal-sourced food than is healthy. Plant-rich diets can help reduce chronic illnesses, such as heart disease, stroke, diabetes, and cancer.

Shop Local and Buy Sustainable

To reduce your food's carbon footprint, buy local and seasonal foods. You'll be helping small businesses and farms in your area and reducing fossil fuel emissions associated with transport and cold chain storage. Sustainable agriculture uses up to 56 percent less energy, creates 64 percent fewer emissions and allows for greater levels of biodiversity than conventional farming. Go one step further and try growing your own fruit, vegetables, and herbs. You can plant them in a garden, on a balcony, or even on a window sill. Set up a community garden in your neighbourhood to get others involved.

Don't Waste Food

One-third of all food produced is either lost or wasted. According to UNEP's Food Waste Index Report 2021, people globally waste 1 billion tonnes of food each year, which accounts for around 8 to 10 percent of global greenhouse gas emissions. Avoid waste by only buying what you need. Take advantage of every edible part of the foods you purchase. Measure portion sizes of rice and other staples before cooking them, store food correctly (use your freezer if you have one), be creative with leftovers, share extras with your friends and neighbours, and contribute to a local food-sharing scheme. Make compost out of inedible remnants and use it to fertilize your garden. Composting is one of the best options for managing organic waste while also reducing environmental impacts.

Dress (Climate) Smart

The fashion industry accounts for 8 to 10 percent of global carbon emissions—more than all international flights and maritime shipping combined—and "fast fashion" has created a throwaway culture that sees clothes quickly end up in landfills. But we can change this. Buy fewer new clothes and wear them longer. Seek out sustainable labels and use rental services for special occasions rather than buying new items that will only be worn once. Recycle pre-loved clothes and repair when necessary.

Plant Trees

Every year approximately 12 million hectares of forest are destroyed and this deforestation, together with agriculture and other land use changes, is responsible for roughly 25 percent of global greenhouse gas emissions. We can all play a part in reversing this trend by planting trees, either individually or as part of a collective. For example, the Plant-for-the-Planet initiative allows people to sponsor tree-planting around the world.

Focus on Planet-Friendly Investments

Individuals can also spur change through their savings and investments by choosing financial institutions that do not invest in carbon-polluting industries. #ActNow Speak Up has a section on money and so does Count Us In. This sends a clear signal to the market and already many financial institutions are offering more ethical investments, allowing you to use your money to support causes you believe in and avoid those you don't. You can ask your financial institution about their responsible banking policies and find out how they rank in independent research.

Viewpoint 5

> *"We need to quickly move to shut down fossil fuel use altogether, by replacing existing infrastructure as quickly as possible with solar, wind, and other non-fossil energy sources so that people can thrive in a safe climate future."*

Ending the Climate Crisis Requires Discontinuing the Use of Fossil Fuels

H. Damon Matthews and Eric Galbraith

In this viewpoint, H. Damon Matthews and Eric Galbraith believe that it is essential to stop burning coal, oil, and natural gas in order to help reduce the impacts of climate change. While other solutions may help, they will not be effective if the use of fossil fuels continues. Global temperatures will continue to rise if more carbon dioxide is emitted into the atmosphere, which will lead to more devastating weather events and permanent changes to the climate. The authors concede that making this change will be a big undertaking, but that the problem itself is simple: burning fossil fuels causes climate change, so we must stop using fossil fuels. H. Damon Matthews is a professor and climate scientist in the department of geography, planning, and environment at Concordia University. Eric Galbraith is a professor of

"Ending the climate crisis has one simple solution: Stop using fossil fuels", by H. Damon Matthews and Eric Galbraith, The Conversation, 11/14/2022. https://theconversation.com/ending-the-climate-crisis-has-one-simple-solution-stop-using-fossil- fuels-194489.

Earth science and Canada Research Chair in Human-Earth System Dynamics at McGill University.

As you read, consider the following questions:

1. What impact do the authors think other solutions to climate change will have if the shift away from fossil fuels doesn't occur?
2. Why do the authors think the climate problem has been presented as more complex than it actually is?
3. What do the authors suggest must be done to end the climate crisis?

As COP27 unfolds in Egypt, we are hearing about a large array of climate change solutions, ranging from building with carbon-absorbing bamboo and using less plastic to growing more kelp in the oceans to retain its carbon stores and enhance biodiversity.

All of these ideas are important and could lead to positive environmental outcomes if successfully implemented.

However, as climate scientists we believe that they also have the potential to be dangerous distractions, drawing attention away from the three things we absolutely need to do to end the climate crisis: Stop burning coal, stop burning oil, and stop burning natural gas.

Ending fossil fuel use is essential to end the climate crisis, and there is no alternative.

A Simple Problem with a Simple Solution

Global warming is fundamentally a very simple problem.

Human use of fossil fuels—whether in the form of coal, oil, or natural gas—releases carbon dioxide (CO_2) and other greenhouse gases into the atmosphere, making Earth's greenhouse effect stronger and increasing Earth's temperature.

The temperature rise, which has reached 1.25 C and counting will continue unless we stop adding CO_2 to the atmosphere. The only way to achieve this is to end the use of coal, oil, and natural gas.

Other solutions including less deforestation, more conserved and restored natural areas, and better agricultural practices could help to slow global warming. These would also have additional benefits such as strengthening biodiversity and increasing community resilience to climate impacts.

Capturing and storing carbon may also eventually play a minor role in limiting the worst effects of climate change, but after decades of research we still don't have a cost-effective strategy to put hundreds of billions of tons of CO_2 back into the ground once the fossil fuels have been burned.

If we don't end the use of fossil fuels, all of the rest adds up to little more than branches piled on the tracks in front of a runaway train. They might slow the train temporarily, but until we get inside the engine and shut off the throttle, the train will keep accelerating.

It's a Big Challenge, but Not a Complex One

The solution to the climate crisis is not complex. But it is big.

The amount of coal, oil, and natural gas extracted from the Earth adds up to billions of tons per year. So, shifting from fossil fuels to improved energy systems will require efforts far beyond what has so far been allocated to solving the climate problem.

But "big" is not the same as "complex." The term "complex" implies that we don't really know what the solutions are or whether they will in fact work. Neither is true here: we know with absolute certainty that replacing fossil fuels with carbon-free energy systems would solve the climate problem. What's more, it's the only way to do so.

So far, we have failed to take the actions necessary to tackle this big, but simple, problem mainly because the problem has been made to appear complex.

Complexifying the nature of the climate problem has been a strategy of the fossil fuel industry for decades, and fossil fuel lobbyists continue to push their agenda at COP27.

For many years, the argument was that the climate system is complex and that maybe greenhouse gas emissions were not a problem. Today, this same guise of complexity is being used to propagate doubt on the efficacy of real solutions and to promote actions that will take us in the wrong direction. Even now, gas representatives at COP27 are suggesting that burning more natural gas is part of the solution.

It is time we looked through this facade of complexity and get to work on the solutions that we know will work.

Fossil Fuel Non-Proliferation Treaty Shows the Way

We do know what needs to be done. The Fossil Fuel Non-Proliferation Treaty is a promising step in the right direction. The treaty, which was launched at the New York Climate Action Week in 2020, called, at last, to stop the expansion of new fossil fuel development and manage a just transition away from coal, oil, and gas to clean energy.

But we can't stop there. We need to quickly move to shut down fossil fuel use altogether, by replacing existing infrastructure as quickly as possible with solar, wind, and other non-fossil energy sources so that people can thrive in a safe climate future.

It's time to look past the distractions and focus on the simple solution to the climate crisis. We need to stop burning coal, oil, and natural gas. Our climate future depends on this.

Periodical and Internet Sources Bibliography

The following articles have been selected to supplement the diverse views presented in this chapter.

Jolene Creighton, "Renewable Energy is Great, But It Won't Save Us," *Futurism*, January 15, 2018. https://futurism.com/renewable-energy-is-great-but-it-cant-save-our-biosphere.

Mike De Socio, "Why Solar Panels Will Likely Keep Getting Cheaper," CNET, December 19, 2023. https://www.cnet.com/home/energy-and-utilities/why-the-cost-of-solar-panels-will-likely-keep-falling/.

António Guterres, "The World Is Burning. We Need a Renewables Revolution," *Africa Renewal*, June 27, 2022. https://www.un.org/africarenewal/magazine/june-2022/world-burning-we-need-renewables-revolution.

Julia Jacobo and David Miller, "Batteries, Solar, Wind and Hydropower: Why Renewable Energy Is Essential to Curbing Climate Change," ABC News, April 20, 2022. https://abcnews.go.com/Business/batteries-solar-wind-hydropower-renewable-energy-essential-curbing/story?id=84019384.

Amory B. Lovings and M. V. Ramana, "Three Myths About Renewable Energy and the Grid, Debunked," Yale Environment 360, December 9, 2021. https://e360.yale.edu/features/three-myths-about-renewable-energy-and-the-grid-debunked.

Michael Lynch, "Renewables Are Not the Answer to the Current Energy Crisis," *Forbes*, March 7, 2022. https://www.forbes.com/sites/michaellynch/2022/03/07/renewables-are-not-the-answer-to-the-current-energy-crisis/?sh=2a049a0f1513.

Christina Nunez, "Are There Real Ways to Fight Climate Change? Yes." *National Geographic*, May 17, 2023. https://www.nationalgeographic.com/environment/article/global-warming-solutions?loggedin=true&rnd=1708811113362.

Sammy Roth, "Can Rooftop Solar Alone Solve Climate Change? Here's the Answer," *Los Angeles Times*, June 29, 2023. https://www.latimes.com/environment/newsletter/2023-06-29/can-rooftop-solar-alone-solve-climate-change-heres-the-answer-boiling-point.

Adie Tomer, Joseph W. Kane, and Caroline George, "How Renewable Energy Jobs Can Uplift Fossil Fuel Communities and Remake Climate Politics," Brookings, February 23, 2021. https://www.brookings.edu/articles/how-renewable-energy-jobs-can-uplift-fossil-fuel-communities-and-remake-climate-politics/.

Jack Yang, "The Energy Bundle Theory – Why Renewable Energy is not the Panacea for Global Inequality," *Berkeley Economic Review*, April 18, 2023. https://econreview.berkeley.edu/the-energy-bundle-theory-why-renewable-energy-is-not-the-panacea-for-global-inequality/.

OPPOSING VIEWPOINTS® SERIES

Chapter 3

Is the Shift to Renewable and Alternative Energy Sources Possible?

Chapter Preface

The world has been running on fossil fuels for so long that they can seem inexhaustible. Gas, coal, and oil seem to have always been here and sometimes seem as if they always will be. But that is far from true. Fossil fuels are a limited resource. Even if burning fossil fuels weren't destroying our environment, humans would eventually have to come up with alternative sources of energy. When we do find new sources of energy, they need to be both renewable and sustainable. We've learned our lessons, and must not make the same mistakes again, say the authors in this chapter.

However, the road to renewable, sustainable sources of energy is not proving to be a smooth one. While the authors in this chapter (and in other chapters) point to plenty of successes, they also point out that we are not moving fast enough. One viewpoint in this chapter examines the odd fact that while renewable energy is now cheaper than fossil fuels, the use of fossil fuels is still growing.

But the situation is far from hopeless. One viewpoint discusses research that shows the vast majority of people in the U.S. (one of the top three emitters of carbon dioxide) support the move toward renewable energy. Answers to the survey referenced in the viewpoint varied by party affiliation, but the lines weren't as starkly drawn as one might imagine. Democrats and Republicans vary in what sorts of policies and changes they're willing to support. But overall, addressing climate change seems to be a winning approach —especially with younger people, as the last viewpoint shows.

Ending on a note that is both optimistic and inspiring, the final viewpoint profiles 10 young African climate activists who have taken up the challenge to fight climate change to protect the people in their countries and around the world.

Viewpoint 1

> *"Rapidly increasing renewable energy hasn't cut coal and gas consumption at the same rate because humankind is using a lot more electricity than we used to."*

Renewables Are Cheaper, Yet Fossil Fuel Use Is Still Growing

Malte Jansen

In this viewpoint, Malte Jansen explains that even though fossil fuels are limited and renewables are now cheaper, the use of fossil fuels is still growing. This can largely be credited to the fact that even though many Western countries have been reducing the amount of coal and gas used for energy generation, these resources are still a major part of energy power systems in other parts of the world, particularly Asia. In part, that is because the demand for energy in these regions currently exceeds what can be produced using wind and solar energy. Malte Jansen is a lecturer in energy and sustainability and an expert in energy policy, renewable power plant technology, and energy market design. He works at the University of Sussex in the UK.

"Renewables are cheaper than ever yet fossil fuel use is still growing – here's why," by Malte Jansen, The Conversation, September 19, 2023, https://theconversation.com/renewables-are-cheaper- than-ever-yet-fossil-fuel-use-is-still-growing-heres-why-213428.

As you read, consider the following questions:

1. Which type of fossil fuel is being phased out most quickly in Europe and North America, according to this viewpoint?
2. Why is coal not being phased out as quickly in Asia?
3. What good news and what bad news does Jansen close with?

Wind and solar are the world's fastest growing energy sources and together generated 12 percent of global electricity in 2023. The amount of energy produced by wind and solar is expected to increase and accelerate.

Wind generated 1 terawatt (TW) for the first time in 2023 —nearly as much as the total installed energy capacity of the United States (1.2 TW). Solar broke this threshold in 2022.

So why in the first global stocktake of the world's progress towards limiting warming to 1.5°C did the UN say we're still not phasing out fossil fuels fast enough?

Asia's Economic Growth Powered by Coal

Despite the rapid growth of renewable energy, the most carbon-intensive forms of electricity generation, using coal and natural gas, have risen by 22 percent and 37 percent since 2010, respectively. Coal and gas power generation is still the backbone of global energy systems and these fuels are likely to remain dominant for decades to come. Nonetheless, the phase-out of coal (arguably the dirtiest of fossil fuels) is gaining momentum.

During the past decade, the number of new coal power plants built each year has fallen fast. Global coal demand has continued to fall even as the war in Ukraine strained gas supplies. In the most prosperous OECD countries (or Organisation for Economic Co-operation and Development), virtually no new coal plants are planned or being built, though new coal mines are still being approved. This is a result of national policies such as the UK's decision to ban coal in power generation from October 2024.

The United States has retired many aging coal plants since the mid-2010s due to the low price of shale gas. The country's coal fleet will continue to shrink as 99 percent of coal projects are more expensive than new clean energy, thanks to the Inflation Reduction Act (IRA).

This picture is very different in Asia. Here, countries have relied heavily on cheap coal to fuel their economies. This is particularly true in China. After adding 27 gigawatts (GW) from coal in 2022 alone, China by itself is offsetting the retirement of coal plants elsewhere in the world.

But there are some signs this is changing. The global pipeline for new coal power plants is smaller than ever and China and India both pledged to "phase down coal" in 2021 at the Glasgow climate summit.

So, rapidly increasing renewable energy hasn't cut coal and gas consumption at the same rate because humankind is using a lot more electricity than we used to, especially in Asia. In the last 20 years, the use of electricity in Europe and North America has remained largely constant. Here, renewable energy has slowly eaten into the proportion of energy generated by fossil fuels, while all other energy sources (nuclear, hydro, biomass) have remained about the same. In Asia, electricity demand has tripled since the 2000s, with the bulk of this energy coming from fossil fuels.

Wind and Solar Are Replacing Coal and Gas

Western economies have made progress in replacing fossil fuels (and coal in particular) with renewables during the last decade. In Europe and North America, wind has become a vital energy source during the winter months when energy demand peaks. And when the wind isn't blowing, gas generation fills the gaps.

Solar energy, when combined with batteries which can store excess electricity, is also proving to be a cheaper option than both gas and coal in certain parts of the world. In Australia, the industry association Australian Clean Energy Council found that solar panels and batteries are 30 percent cheaper than gas power plants during peak demand periods.

A Bloomberg NEF investigation found that batteries alone are already cheaper than gas power plants during these times. In fact, solar panels may be generating electricity more cheaply than the grid in some cases.

In India, the cost of generating electricity from solar and storing it in batteries to use during high demand hours has lower costs than existing coal plants. Combined solar and battery plants can activate during peak hours and turn off again when demand drops, regardless of whether the wind is blowing or the sun shining.

In the United States, almost half of new energy projects waiting to connect to the grid combine solar and wind with storage technologies, allowing renewables to produce electricity on demand regardless of the weather.

Energy Demand Is Outpacing Wind and Solar

Wind and solar has only slowed the rise in fossil fuel burning. This is particularly true for China, India, Thailand, and Vietnam. These economies have grown rapidly and so has their power demand.

The replacement with renewables in developed economies is too slow to offset this increase on a global scale. Cooling economic activity in Asia—especially China—might reverse this trend, making a replacement pattern similar to Europe and North America feasible.

While the warnings in the UN's stocktake should be heeded, the outlook is not entirely gloomy. According to the International Energy Agency (IEA) in a report from December 2022, virtually all new demand between now and 2025 will be satisfied by renewable energy. Wind and solar are expected to supply the bulk of this additional electricity, owing to their low cost and high availability.

With new wind and solar now cheaper than existing fossil fuel generation, it is only a matter of time before they fully replace all new energy demand first, and replace existing fossil fuels after—even in fast-growing economies. However, as the UN report shows, this process needs to be significantly sped up to avert catastrophic warming.

Viewpoint 2

> *"Combating climate change is not cheap. Actions by governments and the private sector are both needed to slow global warming and keep it from fueling increasingly devastating disasters."*

The Transition to Renewable Energy Will Not Be Cheap, and Treaties Favoring Fossil Fuel Investors Make It More Expensive

Rachel Thrasher, Blake Alexander Simmons, and Kyla Tienhaara

In this viewpoint, the authors describe the ways in which fossil fuel companies use treaties to discourage countries from transitioning away from fossil fuels to renewable energy. These treaties were meant to entice fossil fuel companies to bring projects into countries, which would then offer more jobs and new technology to these countries. In exchange, the treaties allow investors to sue governments if projects are cancelled or downsized. Since this could result in a substantial amount of money going toward fossil fuel investors, there will be less money available for the transition to renewable energy. Rachel Thrasher is a law lecturer and researcher at the Boston University Global Development Policy Center. Blake Alexander Simmons is an environmental social scientist for the Tampa Bay Estuary Program and an instructor and teaching assistant for the Harvard Extension School. Kyla Tienhaara is a Canada Research Chair in Economy and Environment and an Assistant Professor in the School of Environmental Studies and the department of global development studies at Queen's University, Canada.

As you read, consider the following questions:

1. What is ISDS? How does it factor into the transition away from fossil fuels?
2. Which countries would be most impacted by fossil fuel industries taking legal action if they violate treaties?
3. What options do countries have to avoid legal and financial risks, according to this viewpoint?

Fossil fuel companies have access to an obscure legal tool that could jeopardize worldwide efforts to protect the climate, and they're starting to use it. The result could cost countries that press ahead with those efforts billions of dollars.

Over the past 50 years, countries have signed thousands of treaties that protect foreign investors from government actions. These treaties are like contracts between national governments, meant to entice investors to bring in projects with the promise of local jobs and access to new technologies.

But now, as countries try to phase out fossil fuels to slow climate change, these agreements could leave the public facing overwhelming legal and financial risks.

The treaties allow investors to sue governments for compensation in a process called investor-state dispute settlement, or ISDS. In short, investors could use ISDS clauses to demand compensation in response to government actions to limit fossil fuels, such as canceling pipelines and denying drilling permits. For example, TC Energy, a Canadian company, is currently seeking more than

US$15 billion over U.S. President Joe Biden's cancellation of the Keystone XL Pipeline.

In a study published May 5, 2022, in the journal *Science*, we estimate that countries would face up to $340 billion in legal and financial risks for canceling fossil fuel projects that are subject to treaties with ISDS clauses.

That's more than countries worldwide put into climate adaptation and mitigation measures combined in fiscal year 2019, and it doesn't include the risks of phasing out coal investments or canceling fossil fuel infrastructure projects, like pipelines and liquefied natural gas terminals. It means that money countries might otherwise spend to build a low-carbon future could instead go to the very industries that have knowingly been fueling climate change, severely jeopardizing countries' capacity to propel the green energy transition forward.

Massive Potential Payouts

Of the world's 55,206 upstream oil and gas projects that are in the early stages of development, we identified 10,506 projects—19 percent of the total—that were protected by 334 treaties providing access to ISDS.

That number could be much higher. We could only identify the headquarters of project owners, not the overall corporate structures of the investments, due to limited data. We also know that law firms are advising clients in the industry to structure investments to ensure access to ISDS, through processes such as using subsidiaries in countries with treaty protections.

Depending upon future oil and gas prices, we found that the total net present value of those projects is expected to reach $60 billion to $234 billion. If countries cancel these protected projects, foreign investors could sue for financial compensation in line with these valuations.

Doing so would put several low- and middle-income countries at severe risk. Mozambique, Guyana, and Venezuela could each face over $20 billion in potential losses from ISDS claims.

If countries also cancel oil and gas projects that are further along in development but are not yet producing, they face more risk. We found that 12 percent of those projects worldwide are protected by investment treaties, and their investors could sue for $32 billion to $106 billion.

Canceling approved projects could prove exceptionally risky for countries like Kazakhstan, which could lose $6 billion to $18 billion, and Indonesia, with $3 billion to $4 billion at risk.

Canceling coal investments or fossil fuel infrastructure projects, like pipelines and liquefied natural gas terminals, could lead to even more claims.

Countries Already Feel Regulatory Chill

There have been at least 231 ISDS cases involving fossil fuels so far. Just the threat of massive payouts to investors could cause many countries to delay climate mitigation policies, causing a so-called "regulatory chill."

Both Denmark and New Zealand, for example, seem to have designed their fossil fuel phaseout plans specifically to minimize their exposure to ISDS. Some climate policy experts have suggested that Denmark may have chosen 2050 as the end date for oil and gas extraction to avoid disputes with existing exploration license holders.

New Zealand banned all new offshore oil exploration in 2018 but did not cancel any existing contracts. The climate minister acknowledged that a more aggressive plan "would have run afoul of investor-state settlements." France revised a draft law banning fossil fuel extraction by 2040 and allowing the renewal of oil exploitation permits after the Canadian company Vermilion threatened to launch an ISDS case.

Securing the Green Energy Transition

While these findings are alarming, countries have options to avoid onerous legal and financial risks.

What Is the Difference Between Renewable and Sustainable Energy?

As the general public becomes more conscientious of the need to mitigate climate change by transitioning to new forms of energy, terms can often be misunderstood or mistakenly used interchangeably. One example is "sustainable energy" versus "renewable energy." While both sustainable energy and renewable energy are critical to the future of our planet, there are subtle distinctions between the two categories. Not all renewable energy is sustainable, and not all sustainable energy sources are renewable.

What Is Renewable Energy?

The definition of renewable energy is usable energy created through naturally recurrent processes—the wind blowing or the sun shining, for example. Some common examples of renewable energy sources are solar, wind, hydro, tidal, geothermal and biomass. Being weather- and time-dependent, these categories are not always readily available, but they will reliably and naturally replenish themselves, making them an ideal and potentially infinite alternative to limited, harmful resources like fossil fuels. Traditional and non-renewable energy sources like coal and oil, which since the industrial age have been used to "power" much of our modern lives, took millions of years to form, and their rate of consumption far outpaces their rate of replenishment.

Renewables are a frequent topic of public discourse and have filtered down to the general public; one 2017 survey indicated that 82 percent of people around the world think the planet should be fueled by renewable energy sources. Notably, renewables are also increasingly becoming more cost-effective, contributing to their growing popularity. But excessive use of even renewable resources can render a particular energy source unsustainable.

What Is Sustainable Energy?

The definition of sustainable energy is an energy resource that can be maintained for the foreseeable future without compromising or threatening future generations. Energy is "sustainable" when it can meet a company or a society's needs without being depleted or becoming unusable in some fashion. Geothermal, hydropower, solar and wind are common sustainable energy sources. Some policy makers and arbiters of public opinion argue that oil, natural gas, and coal are sustainable forms of energy because of their wide availability, but this is controversial and contested as a hard-and-fast stance. Generally, though, energy industry experts can agree that some non-renewable sources of energy can be sustainable if used in moderation.

[...]

"What is the difference between renewable and sustainable energy?," Insight Prysmian Group Magazine.

The Organization for Economic Cooperation and Development is currently discussing proposals on the future of investment treaties.

A straightforward approach would be for countries to terminate or withdraw from these treaties. Some officials have expressed concern about unforeseen impacts of unilaterally terminating investment treaties, but other countries have already done so, with few or no real economic consequences.

For more complex trade agreements, countries can negotiate to remove ISDS provisions, as the United States and Canada did when they replaced the North American Free Trade Agreement with the United States-Mexico-Canada Agreement.

Additional challenges stem from "sunset clauses" that bind countries for a decade or more after they have withdrawn from some treaties. Such is the case for Italy, which withdrew from the Energy Charter Treaty in 2016. It is currently stuck in an ongoing ISDS case initiated by the U.K. company Rockhopper over a ban on coastal oil drilling.

The Energy Charter Treaty, a special investment agreement covering the energy sector, emerged as the greatest single contributor to global ISDS risks in our dataset. Many European countries are currently considering whether to leave the treaty and how to avoid the same fate as Italy. If all country parties to a treaty can agree together to withdraw, they could collectively sidestep the sunset clause through mutual agreement.

The Global Transition

Combating climate change is not cheap. Actions by governments and the private sector are both needed to slow global warming and keep it from fueling increasingly devastating disasters.

In the end, the question is who will pay—and be paid —in the global energy transition. We believe that, at the very least, it would be counterproductive to divert critical public finance from essential mitigation and adaptation efforts to the pockets of fossil fuel industry investors whose products caused the problem in the first place.

VIEWPOINT 3

> *"The findings illustrate ongoing public reluctance to make sweeping changes to American life to cut carbon emissions."*

Americans Support Renewable Energy, but Aren't Willing to Make Big Changes

Brian Kennedy, Cary Funk, and Alec Tyson

Several of the previous viewpoints have discussed the fact that renewable energy, though cheaper and more sustainable, is not being implemented fast enough. In this viewpoint, researchers at the Pew Research Center dig into the data on how Americans feel about renewable energy—and what they're willing to do to help save the planet. In general, Americans support making changes to support renewable energy, but are wary of major changes, such as ending the production of gas-powered vehicles in the near future. Cary Funk is a research consultant and former researcher at Pew Research Center. Brian Kennedy is a senior researcher at Pew. Alec Tyson is Pew's associate director of research.

As you read, consider the following questions:

1. How do Democrats and Republicans differ on their support for efforts to address climate change?

"Majorities of Americans Prioritize Renewable Energy, Back Steps to Address Climate Change," by Brian Kennedy, Cary Funk, and Alec Tyson, Pew Research Center, June 28, 2023.

2. What demographic of the Republican party is most supportive of renewable energy?
3. According to this viewpoint, how do Americans of both parties view President Biden's climate agenda?

A new Pew Research Center survey finds large shares of Americans support the United States taking steps to address global climate change and back an energy landscape that prioritizes renewable sources like wind and solar. At the same time, the findings illustrate ongoing public reluctance to make sweeping changes to American life to cut carbon emissions. Most Americans *oppose* ending the production of gas-powered vehicles by 2035 and there's limited support for steps like eliminating gas lines from new buildings.

This report comes about a year after the 2022 Inflation Reduction Act introduced policies and incentives meant to dramatically reduce the country's reliance on fossil fuels, a signature part of the Biden administration's efforts on climate change. The survey takes stock of how Americans feel about related questions on climate, energy, and environmental policy, including proposed changes to how Americans power their homes and cars and what to do about the impacts communities face from extreme weather.

The Pew Research Center survey of 10,329 U.S. adults conducted May 30 to June 4, 2023, finds:

- 74 percent of Americans say they support the country's participation in international efforts to reduce the effects of climate change.
- 67 percent of U.S. adults prioritize the development of alternative energy sources such as wind, solar, and hydrogen power over increasing the production of fossil fuel energy sources.

By sizable margins, Americans support a number of specific policy proposals aimed at reducing the effects of climate change through targeting greenhouse gas emissions and carbon in the atmosphere:

- Overwhelming majorities support planting about a trillion trees around the world to absorb carbon emissions (89 percent) and requiring oil and gas companies to seal methane gas leaks from oil wells (85 percent).
- 76 percent favor providing a tax credit to businesses that develop carbon capture technologies and 70 percent support taxing corporations based on their carbon emissions.
- 61 percent favor requiring power plants to eliminate all carbon emissions by the year 2040.

Still, there are limits to public support for major changes to the way homes, cars, and the electrical grid are powered.

Only 31 percent of Americans currently support phasing out the use of fossil fuel energy sources altogether. Another 32 percent say the U.S. should eventually stop using fossil fuels, but don't believe the country is ready now. And 35 percent think the U.S. should *never* stop using fossil fuels to meet its energy needs.

Less than half of the public (40 percent) favors phasing out the production of gas-powered cars and trucks. Support for this policy is 7 percentage points lower than it was two years ago. And underscoring the strong feelings big changes to American life can engender, 45 percent say they would feel *upset* if gas-powered cars were phased out; fewer than half as many (21 percent) would feel *excited*.

When it comes to the construction of new buildings, slightly more Americans oppose (51 percent) than favor (46 percent) requiring most new buildings to run only on electricity, with no gas lines, a recent flashpoint in state legislatures and national climate conversations.

Republicans and Democrats continue to offer competing visions on climate and energy issues. Deep Republican skepticism toward a renewable energy transition is a major factor behind much of the overall public's reluctance to make a sharp break from fossil fuels.

But views within both party coalitions defy simple categorization. And some of the most far-reaching policies aimed at addressing climate change and carbon emissions garner a less-than-enthusiastic response from Democrats, as well as outright opposition from Republicans.

A Closer Look at the Two Major Party Coalitions on Climate and Energy Issues

Republicans and Republican-leaning independents prioritize oil, coal, and natural gas development over renewable energy sources and have deep concerns (especially around prices) about what a transition to renewable energy would mean for the country.

Some proposed changes, like phasing out new gas-power vehicles provoke a strong negative response: 73 percent of Republicans say they would be upset by this. And stacked up against other national issues, climate change consistently ranks as a low priority for Republicans.

But these attitudes do not preclude Republican support for climate policies and renewable energy altogether. Two-thirds favor a business tax credit for the development of carbon capture technologies and majorities support more solar panel and wind farms, when these energy sources are not placed in competition with fossil fuel development.

Within the GOP, moderates and young Republicans often offer the most support for action on climate change and a shift toward renewable energy, though they make up a relatively smaller share of all Republicans and GOP leaners compared with conservatives and older Republicans.

Democrats and Democratic-leaning independents overwhelmingly back the U.S. participating in international efforts to address global climate change. By a 90 percent to 10 percent margin, Democrats say renewable energy sources should be given priority over the development of oil, coal, and natural gas.

Democrats, by and large, foresee benefits from an energy transition in the U.S., including better air and water quality, job opportunities in the energy sector, and greater energy independence.

Despite this favorable stance toward climate action and renewable energy, 51 percent of Democrats oppose phasing out fossil fuels altogether, saying instead, that oil, coal, and natural gas should continue to be part of the mix of energy sources the country relies on.

Within the Democratic Party, large majorities across age and ideological groups are generally supportive of shifting toward renewable energy and policies to address climate change.

Still, important differences do emerge, especially regarding the pace of an energy transition: 61 percent of liberals are ready to phase out the use of fossil fuels altogether, while 62 percent of moderates and conservatives say they should be part of a mix of sources for now, along with renewables. And while a large share of liberals (76 percent back ending the production of new gas-powered vehicles by 2035, support among moderates and conservative is more limited (53 percent).

What Role Should the Government Play in Communities at High Risk of Extreme Weather

As communities across the country confront the risks posed by extreme weather, such as wildfires, severe storms, and coastal erosion, Americans express an openness to some policy proposals for communities at high risk from these events.

- More say it's a good idea than a bad idea for the federal government to limit new construction in communities at high risk of extreme weather (54 percent to 19 percent). Another 27 percent say they are not sure of their views.
- 53 percent of Americans think it's a good idea for the federal government to provide financial assistance to help communities rebuild after extreme weather events. Far fewer (24 percent) call this a bad idea and 23 percent are not sure.
- By a 47 percent to 24 percent margin, more say it's a good idea than a bad idea for the federal government to provide financial assistance for communities to relocate from areas at high risk of extreme weather.

- But there's less public support for the federal government *requiring* communities to relocate from high-risk areas: 29 percent call this a good idea, compared with 33 percent who say it is a bad idea, and 38 percent who say they're not sure.

When it comes to firsthand experiences with extreme weather and its connection with climate change:

- 69 percent of Americans say they've experienced at least one of five types of extreme weather in the past year: Long periods of unusually hot weather (45 percent), severe weather such as floods or intense storms (44 percent), droughts or water shortages (33 percent), major wildfires (18 percent), and rising sea levels that erode beaches and shorelines (16 percent).
- Majorities of Americans who say they've experienced extreme weather in the last year believe that climate change has contributed a lot or a little to these events.

How Do Americans See Biden's Climate Agenda Today?

Overall, 45 percent of Americans say the Biden administration's policies on climate change are taking the country in the right direction, while slightly more (50 percent) say they have the country headed in the wrong direction. This overall rating reflects a sharp partisan divide: 76 percent of Democrats see President Joe Biden's climate policies as heading in the right direction; by contrast, 82 percent of Republicans say they're taking the country in the wrong direction.

Within Biden's own party, there are signs that Democrats aren't completely satisfied with the administration's actions on climate. Among the majority of Democrats who think Biden's climate policies are pointed in the right direction:

- 59 percent say that Biden could be doing a lot more on climate change, compared with a smaller share (39 percent) who say he's done as much as can be expected.

- And while 51 percent of Democrats who agree with Biden's overall direction on climate say he's taken about the right approach toward compromise, 34 percent say he has compromised too much on climate policy.

The recently approved Willow oil drilling project in Alaska garnered wide attention in energy and climate circles, but the Biden administration's decision did not register widely with the public: 68 percent of Americans say they've heard nothing at all about this issue; 32 percent say they've heard at least a little about it.

The Biden administration's decision to approve the Willow project is unpopular with Democrats who are aware of it: By 61 percent to 23 percent, more Democrats who have heard of the project oppose than favor it. Liberal Democrats aware of the issue are especially critical (74 percent oppose the decision).

Do Americans Support Government Efforts to Address Environmental Disparities Across Communities?

Most Americans support the federal government playing a major (46 percent) or minor (29 percent) role addressing differences across communities in their health risks from pollution and other environmental problems. A small share (8 percent) say the federal government should play no role in this, while 16 percent say they're not sure.

The issue of environmental health disparities across communities is fairly well known to the public: 77 percent of Americans say they've heard a lot or a little about this issue, while 22 percent say they haven't heard about it.

Large majorities of Democrats (68 percent) and those most familiar with the issue (73 percent) support the federal government playing a *major* role addressing community differences in health risks from pollution and environmental problems.

Viewpoint 4

> *"Renewables are now cheaper in most countries and generate three times more jobs than fossil fuels."*

Renewable Energy Resources Are Plentiful and the Logical Alternative to Fossil Fuels

United Nations

In this viewpoint from the United Nations (UN), the author explains the different types of renewable resources that are available, potential challenges involved in implementing them, and the advantages they offer. There are certain considerations that should be kept in mind when deciding which type of renewable energy would work in a particular region, but the authors argue that it is possible to find a renewable energy source that is replenishable, dependable, and cheaper than fossil fuels around the world. In this sense, the shift to renewable energy is not only possible, but logical. The United Nations is an international organization that is meant to maintain international peace, security, and wellbeing.

As you read, consider the following questions:

1. How does the amount of energy Earth intercepts compare to the amount of energy humans consume?
2. According to this viewpoint, what are the best locations for generating wind power?

3. What is currently the largest source of renewable power for electricity?

Renewable energy is energy derived from natural sources that are replenished at a higher rate than they are consumed. Sunlight and wind, for example, are such sources that are constantly being replenished. Renewable energy sources are plentiful and all around us.

Fossil fuels—coal, oil, and gas—on the other hand, are non-renewable resources that take hundreds of millions of years to form. Fossil fuels, when burned to produce energy, cause harmful greenhouse gas emissions, such as carbon dioxide.

Generating renewable energy creates far lower emissions than burning fossil fuels. Transitioning from fossil fuels, which currently account for the lion's share of emissions, to renewable energy is key to addressing the climate crisis.

Renewables are now cheaper in most countries, and generate three times more jobs than fossil fuels.

Here are a few common sources of renewable energy:

Solar Energy

Solar energy is the most abundant of all energy resources and can even be harnessed in cloudy weather. The rate at which solar energy is intercepted by Earth is about 10,000 times greater than the rate at which humankind consumes energy.

Solar technologies can deliver heat, cooling, natural lighting, electricity, and fuels for a host of applications. Solar technologies convert sunlight into electrical energy either through photovoltaic panels or through mirrors that concentrate solar radiation.

Although not all countries are equally endowed with solar energy, a significant contribution to the energy mix from direct solar energy is possible for every country.

The cost of manufacturing solar panels has plummeted dramatically in the last decade, making them not only affordable but often the cheapest form of electricity. Solar panels have a lifespan of roughly 30 years, and come in variety of shades depending on the type of material used in manufacturing.

Wind Energy

Wind energy harnesses the kinetic energy of moving air by using large wind turbines located on land (onshore) or in sea- or freshwater (offshore). Wind energy has been used for millennia, but onshore and offshore wind energy technologies have evolved over the last few years to maximize the electricity produced with taller turbines and larger rotor diameters.

Though average wind speeds vary considerably by location, the world's technical potential for wind energy exceeds global electricity production, and ample potential exists in most regions of the world to enable significant wind energy deployment.

Many parts of the world have strong wind speeds, but the best locations for generating wind power are sometimes remote ones. Offshore wind power offers tremendous potential.

Geothermal Energy

Geothermal energy utilizes the accessible thermal energy from Earth's interior. Heat is extracted from geothermal reservoirs using wells or other means.

Reservoirs that are naturally sufficiently hot and permeable are called hydrothermal reservoirs, whereas reservoirs that are sufficiently hot but that are improved with hydraulic stimulation are called enhanced geothermal systems.

Once at the surface, fluids of various temperatures can be used to generate electricity. The technology for electricity generation from hydrothermal reservoirs is mature and reliable and has been operating for more than 100 years.

Hydropower

Hydropower harnesses the energy of water moving from higher to lower elevations. It can be generated from reservoirs and rivers. Reservoir hydropower plants rely on stored water in a reservoir, while run-of-river hydropower plants harness energy from the available flow of the river.

Hydropower reservoirs often have multiple uses—providing drinking water, water for irrigation, flood and drought control, and navigation services, as well as energy supply.

Hydropower currently is the largest source of renewable energy in the electricity sector. It relies on generally stable rainfall patterns, and can be negatively impacted by climate-induced droughts or changes to ecosystems which impact rainfall patterns.

The infrastructure needed to create hydropower can also impact on ecosystems in adverse ways. For this reason, many consider small-scale hydro a more environmentally friendly option and especially suitable for communities in remote locations.

Ocean Energy

Ocean energy derives from technologies that use the kinetic and thermal energy of seawater—waves or currents for instance—to produce electricity or heat.

Ocean energy systems are still at an early stage of development, with a number of prototype wave and tidal current devices being explored. The theoretical potential for ocean energy easily exceeds present human energy requirements.

Bioenergy

Bioenergy is produced from a variety of organic materials, called biomass, such as wood, charcoal, dung, and other manures for heat and power production, and agricultural crops for liquid biofuels. Most biomass is used in rural areas for cooking, lighting, and space heating, generally by poorer populations in developing countries.

Modern biomass systems include dedicated crops or trees, residues from agriculture and forestry, and various organic waste streams.

Energy created by burning biomass creates greenhouse gas emissions, but at lower levels than burning fossil fuels like coal, oil, or gas. However, bioenergy should only be used in limited applications, given potential negative environmental impacts related to large-scale increases in forest and bioenergy plantations and resulting deforestation and land-use change.

Viewpoint 5

> *"Understanding why fossil fuels are nonrenewable can help navigate the transition to sustainable energy."*

The Shift to Renewable Energy Is Not Only Possible, but Essential

Jennifer White

In this viewpoint, Jennifer White argues that the switch to renewable energy is not only beneficial to the environment—it is crucial because of the fact that fossil fuels are nonrenewable sources of energy that will eventually run out. White explains the science behind the finite nature of fossil fuels to make the argument for switching to renewable sources. Specifically, she points out that the amount of time needed to create fossil fuels is very large. Furthermore, much of the remaining fossil fuel reserves are difficult and costly to access. Switching to renewable energy is the logical solution for securing the planet's energy future. Jennifer White is a writer who focuses on science, economics, and green energy.

As you read, consider the following questions:

1. How long ago did fossil fuels form?
2. How are fossil fuels formed?
3. According to this viewpoint, how soon might oil reserves run out if there aren't any major discoveries?

"Fossil Fuels' Non-Renewable Status: The Science Behind Their Finiteness," by Jennifer White, EcoFlow, August 22, 2023. Reprinted by permission.

Fossil fuels like coal, oil, and natural gas play an enormous role in the function of modern society. However, their days are numbered. Fossil fuels are nonrenewable, meaning they draw on finite resources and will eventually dwindle once reserves run out. The science behind their formation and availability paints a stark picture of an inevitable future without them.

Understanding why fossil fuels are nonrenewable can help navigate the transition to sustainable energy.

Introduction

Fossil fuels formed hundreds of millions of years ago from dead organic material like plants and animals. This material accumulated and was gradually buried under layers of rock and sediment over vast timescales. The intense pressure and heat turned the organic matter into concentrated forms of carbon we utilize as coal, oil, and natural gas.

This extremely slow formation process means fossil fuels are nonrenewable in any practical human timeframe. New reserves are not being produced anywhere near the rate we are using them. Additionally, fossil fuel deposits are scattered and finite. While improving technology enables access to more reserves, ultimately the supplies will dwindle.

In contrast, renewable energy sources like solar, wind, and hydropower are constantly replenished. The science is clear that fossil fuel formation requires geological timescales we simply don't have. Understanding the nonrenewable nature of fossil fuels makes clear the need to transition to sustainable energy alternatives.

Fossil Fuels Formed Over Millions of Years

The Slow Process of Fossil Fuel Formation

The key aspect that makes fossil fuels nonrenewable is the immensely slow rate of their formation. Coal, oil, and natural gas originates from organic material that lived millions of years ago. Over vast timescales, this material was buried deeper and deeper, exposed to ever-increasing heat and pressure from the surrounding rock and earth.

This extreme pressure and heat very slowly transformed the organic compounds into concentrated hydrocarbons like coal, petroleum oil, and natural gas. The entire process takes place on the order of millions of years or longer—far slower than any timeframe relevant to human societies.

Why This Process Can't Be Replicated

Modern science shows that creating fossil fuel-like compounds requires specific conditions only found deep underground over geological timescales. The intense heat and pressure, combined with anaerobic conditions devoid of oxygen, cannot be artificially generated to any meaningful degree.

Attempting to replicate even a small fraction of the fossil fuel formation process would require mimicking conditions found miles underground. The incredibly slow pace measured in millions of years also cannot be altered or accelerated. That's why fossil fuels are considered nonrenewable—their formation cannot be replicated by humans.

Lengthy Timescales Involved

The sheer timescale involved in fossil fuel formation is central to understanding why they are considered nonrenewable. The slow process unfolding over millions of years contrasts sharply with human timeframes.

For example, detectable coal formation took an estimated 180 million years or more. Oil formation ranges from 20 million years to over 100 million years. Natural gas can form slightly faster but still requires millions of years. The biomass and organic matter containing energy was built up over even longer time periods.

No human wants to wait millions of years for a return on energy investment. And the natural fossil fuel formation process certainly can't keep pace with our current rate of consumption. The science confirms new fossil fuel deposits take far too long to form for any practical renewal.

Limited Supplies in the Earth

Fossil Fuels Found in Finite Deposits

The fossil fuels currently extracted were formed during specific historical periods and geological conditions. This has led to distinct deposits scattered unevenly across the world. Fossil fuel reserves exist in a finite amount of concentrated deposits rather than uniformly dispersed through the earth's crust.

Coal, oil, and gas reserves are nonrenewable because they are constrained to these relatively rare deposits. Coal beds, oil reservoirs, and natural gas basins are limited and localized based on the regional geology and past conditions at the time of formation. Mining extensively consumes these known deposits.

New exploration can uncover more reserves, but they remain finite. The science shows fossil fuels were created in a patchwork fashion, not continuously replenished sources. Their nonrenewable nature stems from an ultimate scarcity in the earth's crust.

Difficulty Accessing Remaining Reserves

While limited fossil fuel deposits remain globally, they are increasingly difficult and costly to extract as current reserves are depleted. Remaining oil reservoirs may be smaller, deeper, or in extreme environments like the Arctic. This raises the costs and energy inputs required for further extraction.

Similarly, surface coal is largely exhausted in many regions, requiring deep mining or tearing up landscapes to access other seams. Shale gas and tar sands require expensive, input-intensive techniques like fracking. As easier to access reserves decline, fossil fuels get harder to renew.

Improving Technology Only Goes So Far

Advances in exploration, extraction, and mining technology have expanded reachable fossil fuel deposits. But they remain bound by geological realities and the laws of physics. Even innovative techniques can't create new reserves or overcome the fundamental difficulty of renewing nonrenewable resources.

Accessible oil reservoirs and gas basins are likely to eventually deplete. Improving technology plays an important role in utilizing remaining resources but does not make fossil fuels renewable. There are ultimately limits and diminishing returns to how far it can stretch nonrenewable reserves.

Renewables Replenish, Fossil Fuels Don't

The Sun, Wind, and Water Constantly Renew

Unlike finite fossil fuels, renewable energy sources are capable of renewable generation within practical human timescales. The core renewable technologies—solar, wind, hydropower, geothermal, and biomass—tap into flows of energy that are continuously replenished.

The sun provides limitless solar energy. Wind is created by atmospheric currents which are constantly renewed. The water cycle endlessly renews hydropower potential. Geothermal derives from Earth's residual heat, which is effectively limitless. Biomass can be replanted on timescales measurable in years or decades.

In contrast to million-year fossil fuel formation, renewables can thus provide a sustainable flow of energy over human timescales. This ability to rapidly renew makes renewable generation fundamentally different.

Fossil Fuels Use Finite Resources

Renewables utilize Earth's dynamic flows of energy and resources. Fossil fuels are the product of a finite stock of transformed organic matter. This organic matter was created in specific eras and regions and will not be replenished once extracted and consumed.

While the resource flows powering renewables will continue for billions of years, fossil fuel reserves are strictly determined by past depositional environments. Renewables have access to effectively infinite energy sources like sunlight and wind, while fossil fuels draw down a finite stock limited by geology.

Estimates Suggest Fossil Fuels Will Run Out

Projections indicate that at current and growing rates of consumption, economically recoverable fossil fuels will be severely depleted by the end of this century. Oil reserves may largely run out within 50 years without major new discoveries. Coal may run out within 150 to 200 years.

Since new formation and discovery happens slowly, while consumption grows exponentially, supplies face an inevitable decline. The renewable flow of solar, wind, and hydro energy can last forever, but the one-time stock of fossil fuels is being rapidly depleted. Their nonrenewable essence means running out is inevitable.

Alternatives Are Needed

Renewables Offer Sustainable Options

Renewable energy technologies like solar PV, onshore/offshore wind, hydropower, and geothermal are already growing rapidly as alternatives to finite fossil fuels. Renewables can provide sustainable, clean energy flows long into the human future.

The intermittency of renewables can be balanced via storage, transmission grids, demand management and alternative flexible sources like bioenergy. Integrated intelligently, renewable energy can replace dwindling fossil fuels and power the world sustainably.

Fossil Fuels Have a Deadline

Replacing fossil fuels is an urgent necessity due to their nonrenewable nature. Reliance on finite resources means they come with an expiration date, likely within this century for oil and gas.

Unabated use would have catastrophic climate impacts as well.

Shift to renewable energy is inevitable. The only question is whether it will happen quickly enough to avoid severe shortages, price spikes, and environmental damages along the way. Their non-renewable essence gives fossil fuels a looming deadline.

Change Is Inevitable as Supplies Dwindle

Global economies will have to undergo massive structural changes as fossil fuel supplies inevitably decline. Entire industries and energy infrastructure will need to be reworked in the coming decades. This complex transition is already underway but must accelerate.

With fossil fuels tightly woven into our economies, true energy security now requires rapid expansion of renewables, not increasing dependence. Renewables free countries from the geopolitical constraints of nonrenewable resources. The conversion to sustainable energy is a generational challenge.

Conclusion

Fossil fuels are fundamentally nonrenewable due their origins deep in geological history and the immense timescales required for their formation. The finite deposits in Earth's crust are being rapidly depleted by growing consumption. Renewable energy flows can sustainably replace them, but the transition will be an unprecedented logistical and infrastructural overhaul. Understanding the science behind why fossil fuels cannot renew reveals the immense scale of the challenge. But it also makes clear that renewables offer the only long-term path to energy security and environmental sustainability. The fossil fuel era is ending, and the renewables era has just begun.

Viewpoint 6

> *"Despite pumping out the least climate-wrecking emissions, Africa is experiencing more rapid global heating than the rest of the world."*

The Next Generation Is Stepping Up to Fight Climate Change and Demand Climate Justice

Angelo Louw

As previous viewpoints have suggested, young people are generally more supportive of attempts to address climate change than are older people. In this viewpoint, Angelo Louw profiles ten young African climate activists. Their cause includes climate justice, since the continent of Africa emits the least carbon but suffers some of the most devastating effects of climate change. By taking action to ensure that more environmentally friendly choices are made going forward, young people can help support a greener future. Angelo Louw is a content editor for Greenpeace International and is based in Johannesburg, South Africa.

As you read, consider the following questions:

1. What kinds of successes are the young people mentioned in this viewpoint having?

"10 African youth climate activists changing the face of the planet," by Angelo Louw, Greenpeace International, October 21, 2021. https://www.greenpeace.org/international/story/50006/10-african-climate-youth-activists-changing-the-face-of-the-planet/.

2. Why do the people profiled here think that it is important for young people to get involved in the climate movement?
3. What — beyond the climate crisis—seems to motivate the people profiled in this viewpoint?

While I may still qualify as a youth in my country (South Africa), the grey hair I've collected for my troubles as a social justice and climate activist will tell you a very different story. Mine has been a decade-long journey, and I am only just beginning to see the impact of this work; yet there is a whole new generation of African climate youth activists who are already making waves with substantially fewer years under their belt.

Despite pumping out the least climate-wrecking emissions, Africa is experiencing more rapid global heating than the rest of the world. Extreme weather events are increasing in frequency across the continent, and damage from rains, droughts, floods, and storms are intensifying.

It is no wonder that these 10 African youth climate activists felt compelled to step up and hold governments and big polluters accountable for the loss and damage they've caused that's affecting all our lives right now, and as well as future generations. One day the world will look back at them as heroes; but for now, allow me to introduce you to the 10 African youth climate activists who are changing the face of the planet.

African Youth Climate Activists You Need to Know

Yero Sarr, Senegal (20)

Yero is a student and cofounder of the Fridays For Future movement in Senegal. He first got involved in the youth climate activist space at the age of 16. Fears around the future of young people across the planet prompted him to become very active in the fight against climate change—and to make sure that others join him too! He believes that collective action is far better than individual action. Yero's strength is mobilising people; he works with several environmental organisations on many issues important to him (including the impacts of harmful industrial fishing on West African communities).

Twitter: @SarrYero2

Raeesah Noor-Mahomed, South Africa (18)

In their final year of high school, Raeesah led a school boycott to demand that the country's environmental department declare a climate emergency. Fast-forward to a year later, and they have been invited as an observer to COP26 (commencing in Glasgow later this month). The Johannesburg-based intersectional activist strives to decolonise Africa, connecting to activists around the continent to make activism more inclusive and accessible with the organisations Stage For Change and 65 Years.

Twitter: @raeesahnm

Ahmed Elhadj Taieb, Tunisia (22)

Tunisia's youngest climate negotiator will be presenting on behalf of his home country during COP26. Ahmed believes that in 30 years, the youth will become the new decision-makers and things will start to change. However, the environmental science student—who is pursuing a master's in environmental engineering—isn't just waiting around for the day to come. He is the General Secretary of Youth for Climate Tunisia and a #breakfreefromplastic youth ambassador. And if that wasn't enough, he is an environmental blogger with the Fibelk science project.

Twitter: @Ahmed Elhadj

Remy Zahiga, Democratic Republic of Congo (24)

Remy is an African climate youth activist living in the Congo Basin, the world's second largest rainforest. He is no stranger to the impacts of environmental degradation, especially with the constant threat that industrial logging poses on the forest and the communities that live there.

The geology graduate is very passionate about climate change because it directly affects each and every one of us. He feels that world leaders are far from taking decisive action on the climate crisis, which is why we all need to get involved in the climate justice movement.

Twitter: @Remy_Zahiga

Dixon Bahandagira, Uganda (23)

This environmental science student has a massive ambition. He wants to plant 1 million trees in his home country, Uganda, to counter the impact of the climate crisis there. He says that more than 200 lives have been lost in his home country since 2017 due to severe floods and landslides exacerbated by climate change. He fears that the situation will get worse due to lack of political will and has decided to tackle the issue head-on. Dixon has already planted more than 100,000 trees in the space of a year, teaching local people to take care of the plants until they are fully grown.

Twitter: @BahandagiraD

Winnie Cheche, Kenya (31)

Winnie is a young conservationist, blogger, climate activist, volunteer, and communication lead at Kenya Environmental Action Network. Growing up in extreme poverty had silenced her for a great part of her life; but, her passion for wildlife and nature helped her find her voice again—one of the loudest voices in the African youth climate movement. Her activism journey started off online with tweets and posts on matters concerning wildlife welfare and climate change issues. Making social media posts grew into writing blogs and thought leadership and eventually handling communication strategies for one of the organisations leading the Kenyan climate justice movement.

Twitter: @WinnieCheche

Gabriel Klaasen, South Africa (23)

My fellow 2021 Mail & Guardian 200 Young South Africans Award finalist has become a very prolific voice in the South African youth climate movement because of their persistence on viewing the climate crisis through an intersectional lens. Gabriel, based in

Cape Town, is the communications officer and youth coordinator at African Climate Alliance. Their powerful thought leadership has forced the climate movement to acknowledge how the interlocking crises of social injustice and climate breakdown impact some of society's most affected people and areas (MAPA)—such as BIPOC (Black, Indigenous, and people of color) and LGBTQ+ (lesbian, gay, bisexual, transgender, queer)—as well as their active role in shaping a better future for us all.

Twitter: @gabriel_klaasen

Fatna Ikrame El Fanne, Morocco (22)

Somewhere between being a hardcore academic—whose expertise range from engineering to language studies—and an accredited English teacher, Fatna somehow managed to find time to cofound the Youth for Climate Morocco movement. She is also working to tackle discrimination in her role as ambassador for Morocco's AFCD Foundation. Her vision for the future entails beating climate change, poverty, and violence, as well as a world where the word racism does not exist.

LinkedIn: Ikrame elfanne

Evelyn Acham, Uganda (30)

When her friend Vanessa Nakate made international news with a one-woman strike in Uganda's capital, Kampala, Evelyn was so moved that she promised to join her on the frontlines. As she learned more about the climate crisis, the African youth climate activist began to understand the intersectionality between climate change and race—and why it was so important to include it in her home country's school syllabus. As the national coordinator of the Rise Up Movement, Evelyn remains at the forefront of the Ugandan climate movement, organizing climate strikes and calling for divestment from fossil fuels.

Twitter: @eve_chantel

Anisa Bek Derna, Libya (24)

Anisa (aka Nissa) Bek is a media personality, an African youth climate activist, and the founder of Project Mulan—a youth-led project that aims to spread and achieve the United Nations Development Plan's sustainable development goals in Libya. In addition, she is the national leader of Let's Do it World and Fridays for Future in her home country. Nissa believes that a prosperous future where no one is left behind is possible, but it will take hard work. Her hard work over the past nine years has not gone unnoticed, and she recently received the 2021 Diana Award.

Twitter: @Nissa_bek

Periodical and Internet Sources Bibliography

The following articles have been selected to supplement the diverse views presented in this chapter.

Paul Bonifas and Tim Considine, “The Limits to Green Energy,” *Regulation*, Winter 2022 - 2023, https://www.cato.org/sites/cato.org/files/2023-01/regulation-v45n4-3.pdf.

Ronald Brownstein, “The Winds Are Changing for Renewable Energy,” the *Atlantic*, July 7, 2016. https://www.theatlantic.com/politics/archive/2016/07/are-the-winds-changing-for-renewable-energy/490250/.

Ed Conway, “The Paradox Holding Back the Clean Energy Revolution,” *New York Times*, February 22, 2024. https://www.nytimes.com/2024/02/22/opinion/vegas-sphere-energy-efficiency.html.

Suriya Jayanti, “Nuclear Power Is the Only Solution,” *Time*, December 4, 2023. https://time.com/6342343/nuclear-energy-climate-change/.

Maxine Joselow, “How dark money groups led Ohio to redefine gas as ‘green energy,’” *Washington Post,* January 17, 2023. https://www.washingtonpost.com/climate-environment/2023/01/17/ohio-natural-gas-green-energy/.

Nina Lakhani, “Is Hydrogen Really a Clean Enough Fuel to Tackle the Climate Crisis?,” the *Guardian*, March 7, 2023. https://www.theguardian.com/environment/2023/mar/07/hydrogen-clean-fuel-climate-crisis-explainer.

Eric Levitz, “Why You Probably Shouldn’t Blow Up a Pipeline,” *Vox*, February 20, 2024. https://www.vox.com/politics/24074408/climate-change-blow-up-pipeline.

Bianca Nogrady, “This Bold Plan to Kick the World’s Coal Habit Might Actually Work,” *Wired*, August 15, 2023. https://www.wired.com/story/this-bold-plan-to-kick-the-worlds-coal-habit-might-actually-work/.

Harry Stevens, “We Need an Area the Size of Texas for Wind and Solar. Here’s How to Halve It,” *Washington Post*, May 10, 2023. https://www.washingtonpost.com/climate-environment/interactive/2023/renewable-energy-land-use-wind-solar/.

Andrew Stuttaford, "Renewable Energy: Intermittency Has Consequences," *National Review*, November 11, 2022. https://www.nationalreview.com/corner/renewable-energy-intermittency-has-consequences/.

Chapter 4

Can Renewable and Alternative Sources Produce Enough Energy for the Future?

Chapter Preface

As the viewpoints throughout this volume have demonstrated, renewable and alternative energy sources can dramatically reduce emissions of greenhouse gases, even if they can't in themselves totally solve the problem. The viewpoints have also indicated that renewable energy sources are becoming cheaper by the day. In many cases renewable energy is already less expensive than fossil fuels. But there's one important question still to be answered: Can renewable and alternative energy sources produce enough energy to power the world now and in the future? That's the question the authors in this chapter take on.

Various players in the energy sector attended a workshop hosted by the World Bank to discuss how to incentivize renewable power so that it can become efficient enough and cost-effective enough to meet the world's energy needs. The first viewpoint in this chapter reports on suggestions that were made at this meeting.

The next viewpoint takes a look at an intriguing model for working out how to design a renewable energy system — a hybrid of wind and solar in this case — that meets energy needs of a given area despite the limitations of weather and periods without sun.

The third viewpoint, from the United Nations, argues that not only is it possible to power the world with renewables, doing so would improve the economics and the health of all nations—in short, switching to renewables would make the world a better place.

One way to make sure renewable sources of energy can meet the demand is to lessen the demand, and one of the biggest demands for energy is for cooling. Global warming is causing plenty of disasters, such as stronger storms, flooding, and droughts. But one of the biggest killers is heat waves. One viewpoint in this chapter looks at changes cities can make to lessen the need for power to cool them.

Viewpoint 1

> *"There was consensus among practitioners that rapid progress is needed on both the supply and demand fronts."*

Increasing the Efficiency of Renewables Will Take Policy Changes

The World Bank Group

If renewable energy is going to power the future, the world must be invested in it, argues the author of this viewpoint. How best to encourage that investment is still a matter of debate. In this viewpoint, various experts attending a World Bank workshop weigh in on the best policies for adopting renewable energy and increasing its efficiency and cost-effectiveness. Finding a way to make renewable energy economically feasible is one of the main ways to support its widespread adoption. The World Bank Group is an international financial institution that provides loans and grants to the governments of low- and middle-income countries.

As you read, consider the following questions:

1. Why are incentives needed to transition to renewables if the cost of renewables is dropping?

"What Policies Work Best to Deploy Renewable Energy and Increase Energy Efficiency?," The World Bank Group, February 15, 2012, https://www.worldbank.org/en/news/feature/2012/02/15/what- policies-work-best-to-deploy-renewable-energy-and-increase-energy-efficiency.

2. There is tension between the need to deliver energy at prices users can afford and the need for companies to get a return on their investment. How could government policies relieve this tension?
3. What is the Standard Offers (SO) mechanism and why is it working in South Africa?

Investment in renewable energy worldwide rose to about $250 billion in 2010, up from $160 billion in 2009, to reach about 20 percent of new investment in the energy space. This is rapid growth, but most analysts agree that it must reach several trillions of dollars a year to propel the needed transition to a sustainable energy future. Most also cite "appropriate policy incentives" as essential to achieve that.

But what policy incentives for renewable energy and energy efficiency work best? How should they be tailored to fit different conditions? How can policy-makers make energy markets respond to deliver clean energy at affordable prices? These were some of the questions addressed at a recent World Bank workshop that produced guidance to policymakers, now captured online for a wider audience.

At the workshop, government and energy utility policymakers compared notes with private sector energy providers and analysts on experience with feed-in tariffs, renewable portfolio standards, and energy auctions, among others.

Cost: The Key Factor for Policymakers

"We are in a phase of developing higher-cost renewable energy, and this is a highly policy-dependent environment," Tim Richards, managing director of energy policy for General Electric (GE), told workshop participants. The cost of electricity, he added, is the key factor on which policymakers should focus.

Cost is indeed the focus of feed-in tariffs (FITs)—the most widespread policy incentive for renewable energy—which offer cost-based compensation to renewable energy producers. But even if the cost of wind and solar power technologies is dropping, workshop participants agreed that they cannot yet compete with fossil fuels, absent policy incentives.

"FiTs are the policy driver in almost every PV (photo-voltaic solar) success story," said Mark Fulton, head of climate change investment at Deutsche Bank.

Sean Whittaker, senior renewable energy specialist in IFC's Climate Business Group, agreed that FiTs can be effective in driving a shift to renewable energy. But, he added, "electricity decisions are political…they may be motivated by a push for climate or environmental gains, but they could also be driven by demands for social development, to create jobs, reduce costs, or to develop a manufacturing base. So the procurement program must address these demands while also providing certainty for developers."

Given these realities, the workshop addressed the challenge of balancing the need to assure a reasonable return to companies for their investments in producing, transmitting, and distributing renewable energy, and the need to deliver electricity at rates consumers can afford.

One way to do this, the workshop was told, is to use energy auctions for price discovery, to establish price benchmarks before bringing in FiTs. But some warn that energy auctions can be "gamed" by bigger players if they are not well designed.

Auctions cannot operate in a vacuum, said Luiz Barroso of PSR Consulting. With adequate regulatory frameworks effectively implemented, he said, auctions can be a powerful mechanism to scale up renewables. In Brazil, for example, prices for 2.9 gigawatts (GW) of wind energy were as low $60 per MWh after reverse auctions in 2011.

Overall, workshop presenters agreed that policy design and precision are essential to get the best deals for renewable energy

Among the key points that emerged from the workshop:

- a package of complementary policy instruments is most likely to deliver a gradual shift to a lower-carbon energy path

- governments need to set a clear policy purpose and vision for renewable energy and energy efficiency that is certain, long term and transparent
- FiTs and auctions' success depends on system and market conditions: legal and regulatory conditions, institutional and administrative efficiency, and the existence and use of risk-mitigating financial instruments
- RE policies must be coordinated with local energy market conditions
- Policies that have proven their value over time include:
    - Reverse auctions
    - Adequate FiTs with long-term power purchase agreements
    - Mandatory access to the grids
    - Incremental cost pass-through

Energy Efficiency: Efforts Needed on Both Supply and Demand

A mix of policy tools is also needed to scale up energy efficiency, the workshop was told. To promote supply-side energy efficiency, the most common incentives are competitive bidding to elicit proposals that offer the greatest kilowatt-hour (kWh) savings, or standard offers in which the government regulator sets the price. Several experts at the workshop lamented the fact that energy efficiency incentives lag behind those for renewable energy.

South Africa is an exception. Tom Skinner, energy efficiency manager at ESKOM, reported encouraging results from the introduction of standard offers (SO). This SO mechanism, analogous to a FiT, pays a pre-determined amount for energy saved, or for load curtailed. South Africa plans to scale up the SO model and is considering introducing competitive procurement of energy efficiency in the future.

These must be complemented by efforts to reduce demand for energy, said Xiaodong Wang, World Bank senior energy specialist in East Asia. “Financial measures can change consumer behavior and mitigate rebound effects (in which gains in efficiency in one area simply prompt greater energy demand in another),” she said. “But these require strong political will for energy pricing reform.”

There was consensus among practitioners that rapid progress is needed on both the supply and demand fronts. To do that, developing countries need support for policy design and financing of energy efficiency as well as renewable energy, on a level-playing-field, based on thorough assessments of their relative attractiveness.

The co-sponsors of the workshop, the World Bank’s Sustainable Energy Department, the Climate Business Group at the International Finance Corporation (IFC), and the Energy Sector Management Assistance Program (ESMAP), offer services and products in all of these areas.

Viewpoint 2

> *"When uncertainties are taken into account, the problem gets a whole lot trickier."*

A Standalone Hybrid System Shows How to Meet Energy Demand with Renewable Energy Sources

Pierre-Louis Poirion

One of the challenges of renewable energy, such as solar and wind power, is that they are not available all the time, and battery storage is not 100 percent efficient. In this viewpoint, Pierre-Louis Poirion looks at an algorithmic model that shows how energy can be used to better balance production and demand in a renewable energy system. Through eliminating some of the uncertainties involved with converting to renewable energy sources, scientists can help encourage governments to make the switch. Pierre-Louis Poirion is a researcher at the University of Paris-Saclay.

As you read, consider the following questions:

1. What problem with renewable energy systems is addressed in this viewpoint?
2. What is meant by a hybrid system?

3. What were the "uncertainties" the researchers had to work into their model?

Relying less on fossil fuels and more on renewable energy is one of the key challenges of energy transition. Large quantities of greenhouse gases are released into the atmosphere when fossil fuels are extracted and transformed into energy (to be used as motor fuel or for heating, for example). The steadily increasing concentration of greenhouse gases in the atmosphere is causing a rise in global temperatures that could have disastrous effects.

The trouble is, renewable energy sources are not as predictable or stable as diesel or nuclear power.

For example, how many solar panels does it take to meet a given demand for electricity? Researchers from ENSTA ParisTech and ENSIIE have found the answer. They have optimized the design of a stand-alone hybrid energy system—stand-alone meaning not connected to the public electricity grid—composed of wind turbines, solar panels, and batteries. Such a system must be able to meet the demand for electricity while minimizing costs, including the cost of the diesel generator that provides backup power if the other energy sources fall short.

How the System Works

When weather conditions are favorable, the wind turbines and solar panels are enough to meet demand and the excess energy is stored in batteries. These batteries are only used if the demand exceeds what the wind turbines and solar panels can supply.

Unfortunately, battery operation results in energy loss; not all the kilowatt-hours stored can be used. As the system is off-grid, a diesel generator is used to satisfy demand once the batteries are empty. But this is expensive—the cost per kilowatt-hour of generator usage is higher than that of a wind turbine.

A few types of wind turbine, solar panel, and batteries were preselected by electricity specialists. The type and quantity of

Switching to Renewable Energy Could Save Trillions

Switching from fossil fuels to renewable energy could save the world as much as $12tn (£10.2tn) by 2050, an Oxford University study says.

The report said it was wrong and pessimistic to claim that moving quickly towards cleaner energy sources was expensive.

Gas prices have soared on mounting concerns over energy supplies.

But the researchers say that going green now makes economic sense because of the falling cost of renewables.

"Even if you're a climate denier, you should be on board with what we're advocating," Prof Doyne Farmer from the Institute for New Economic Thinking at the Oxford Martin School told BBC News.

"Our central conclusion is that we should go full speed ahead with the green energy transition because it's going to save us money," he said.

The report's findings are based on looking at historic price data for renewables and fossil fuels and then modelling how they're likely to change in the future.

The data for fossil fuels goes from 2020 back more than 100 years and shows that after accounting for inflation, and market volatility, the price hasn't changed much.

Renewables have only been around for a few decades, so there's less data. But in that time continual improvements in technology have meant the cost of solar and wind power have fallen rapidly, at a rate approaching 10% a year.

The report's expectation that the price of renewables will continue to fall is based on "probabilistic" modelling, using data on how massive investment and economics of scale have made other similar technologies cheaper.

"Our latest research shows scaling-up key green technologies will continue to drive their costs down, and the faster we go, the more we will save," says Dr Rupert Way, the report's lead author from the Smith School of Enterprise and the Environment.

Wind and solar are already the cheapest option for new power projects, but questions remain over how to best store power and balance the grid when the changes in the weather leads to fall in renewable output.

[...]

"Switching to renewable energy could save trillions – study," by Jonah Fisher, BBC, September 13, 2022.

each component has to be optimally determined to meet a given demand while minimizing the total cost, which includes the cost of investment (land rental and the purchase and installation of the components) and the operating costs (maintenance and diesel).

Taking Weather Variations into Account

This is a complex problem, since uncertainties about demand and production mean that many scenarios are possible. The system must be robust, which means "good enough" whichever scenario plays out. The model has to take into account variations in weather conditions over the year (which was divided into 8,760 periods of one hour). Fortunately, the total variation is limited because the uncertain data can only vary between given bounds.

First, the researchers designed the optimal system without uncertainties. At this stage, the problem is easy to solve because it has only three integer variables (the number of wind turbines, solar panels, and batteries), or nine if different types of materials can be used.

When uncertainties are taken into account, the problem gets a whole lot trickier. To cope with this complexity, the researchers developed a model with two levels. The first corresponds to sizing, or the number of components to be installed. The second, called the "recourse problem," seeks to optimize the system's performance if the worst-case scenario should occur for the system selected at the first level. The researchers had to deal with more than 40,000 variables (including 8,000 integer variables) and 50,000 constraints.

A Custom Algorithm

The scientists developed an algorithm combining branch-and-bound methods with dynamic programming, and which solves the recourse problem much faster than standard algorithms in most cases. They tested it with real data from three locations with different climates: Montana, USA, with a continental climate; an island in the Philippines with a tropical climate; and Dunkirk, France, with an oceanic climate.

In these real cases, the demand and the reference values for wind and solar production are average values over several years for each time period. The scenarios are limited variations around these values. The results show that if the system is capable of meeting demand at certain critical periods (critical in terms of demand or production), it can also cover any other period.

In Dunkirk, the algorithm showed that seven solar panels, 24 wind turbines and 384 batteries could ensure sufficient production capacity all year round. This problem was quick to solve (160 steps) because the climate is much more stable than in the Philippines (6,000 steps required). For the small island in the Philippines, the optimal system consists of 10 solar panels, 67 wind turbines and 105 batteries, for an annual cost of 31,000 euros.

Note that the budget corresponds to the worst-case scenario. The real cost will fall somewhere between this budget and an easily calculated minimum cost reflecting the ideal climate conditions for the system.

This type of stand-alone hybrid energy system is particularly useful on islands or in remote areas. But the algorithm can also be applied to other similar problems, such as inventory management problems.

Viewpoint 3

> *"Accelerating the transition to clean energy is the pathway to a healthy, livable planet today and for generations to come."*

Renewable Energy Will Power a Better World

United Nations

In this viewpoint from the United Nations, the authors list five reasons for rapidly transitioning to renewable energy. These reasons are widely accepted and include the availability of renewable energy sources, the lower price of renewable energy sources, and the positive impacts on health, the environment, and the economy. The underlying theme here is that renewables will not only prevent the worst ravages of climate change, but they will also help build a better world starting now. The United Nations is an international organization intended to maintain international and security and support the wellbeing of all nations.

As you read, consider the following questions:

1. How will shifting to renewable energy help countries diversify their economies?
2. How does investing in renewable energy create jobs?
3. How do the authors of this viewpoint suggest that countries with limited resources can make the change to renewables?

Energy is at the heart of the climate challenge – and key to the solution.

A large chunk of the greenhouse gases that blanket the Earth and trap the sun's heat are generated through energy production, by burning fossil fuels to generate electricity and heat.

Fossil fuels, such as coal, oil and gas, are by far the largest contributor to global climate change, accounting for over 75 percent of global greenhouse gas emissions and nearly 90 percent of all carbon dioxide emissions.

The science is clear: to avoid the worst impacts of climate change, emissions need to be reduced by almost half by 2030 and reach net-zero by 2050.

To achieve this, we need to end our reliance on fossil fuels and invest in alternative sources of energy that are clean, accessible, affordable, sustainable, and reliable.

Renewable energy sources – which are available in abundance all around us, provided by the sun, wind, water, waste, and heat from the Earth – are replenished by nature and emit little to no greenhouse gases or pollutants into the air.

Fossil fuels still account for more than 80 percent of global energy production, but cleaner sources of energy are gaining ground. About 29 percent of electricity currently comes from renewable sources.

Here are five reasons why accelerating the transition to clean energy is the pathway to a healthy, livable planet today and for generations to come.

1. Renewable Energy Sources Are All Around Us

About 80 percent of the global population lives in countries that are net-importers of fossil fuels—that's about 6 billion people who are dependent on fossil fuels from other countries, which makes them vulnerable to geopolitical shocks and crises.

In contrast, renewable energy sources are available in all countries, and their potential is yet to be fully harnessed. The International Renewable Energy Agency (IRENA) estimates that

90 percent of the world's electricity can and should come from renewable energy by 2050.

Renewables offer a way out of import dependency, allowing countries to diversify their economies and protect them from the unpredictable price swings of fossil fuels, while driving inclusive economic growth, new jobs, and poverty alleviation.

2. Renewable Energy Is Cheaper

Renewable energy actually is the cheapest power option in most parts of the world today. Prices for renewable energy technologies are dropping rapidly. The cost of electricity from solar power fell by

India Makes the Shift to Clean Transportation

India is framing policies to promote the use of clean fuels, including electric vehicles (EVs), and tightening emission norms to meet its carbon reduction targets, transport minister Nitin Gadkari told the Reuters Impact conference on Monday. India is the third-largest user of transport automobiles in the world but 70% of its transport energy need is fulfilled by importing fossil fuels. "The aim is to gradually shift to fuels, which are import substitutes, cost effective, indigenous and pollution free," Gadkari said, adding that this includes biofuels, ethanol blends as well as hybrid EVs and hydrogen fuel cells.

The government will also stick firmly to its 2022 deadline for implementing tighter fuel efficiency norms also known as Corporate Average Fuel Efficiency (CAFE), Gadkari said, which could push automakers to adopt cleaner fuels to meet the new target.

"The government is firm on adhering to the CAFE regulation, where automakers are required to keep average CO2 emission less than 130 grams per kilometre till 2022 and below 120 grams per kilometer thereafter," he said.

"The aim is to gradually shift to fuels, which are import substitutes, cost effective, indigenous and pollution free," Gadkari said.

Many Indian automakers including India's largest carmaker Maruti Suzuki have sought delays for implementation of stricter emission rules.

85 percent between 2010 and 2020. Costs of onshore and offshore wind energy fell by 56 percent and 48 percent respectively.

Falling prices make renewable energy more attractive all around – including to low- and middle-income countries, where most of the additional demand for new electricity will come from. With falling costs, there is a real opportunity for much of the new power supply over the coming years to be provided by low-carbon sources.

Cheap electricity from renewable sources could provide 65 percent of the world's total electricity supply by 2030. It could decarbonize 90 percent of the power sector by 2050, massively cutting carbon emissions and helping to mitigate climate change.

> India is targeting reducing carbon emissions by 33%-35% by 2030 as part of its commitment under the Paris Climate Agreement, Gadkari said, and it is looking at sustainable mobility and clean energy to achieve its goal.
>
> Gadkari's comments come ahead of UN Climate Change Conference (COP26) in November seen as a crucial chance to wring out ambitious enough commitments from governments to address global warming.
>
> However, he said India expects new incentives towards climate funds for developing economies from rich nations, echoing similar comments by India's chief economic adviser last week.
>
> Wealthy nations are under ever-greater pressure to deliver on an unmet pledge, made in 2009, to send $100 billion a year to help finance an adequate response by developing countries to rising global temperatures as the world prepares for COP26.
>
> The South Asian nation intends electric cars to make up 30% of total private car sales by 2030 and for electric motorcycles and scooters to make up 40% of total sales, Gadkari said.
>
> The government will also soon make it mandatory for gasoline cars to have flexible fuel engines so they can also run on ethanol blends, Gadkari said, adding that India aims to achieve 20% ethanol-blending with gasoline by 2025 - five years ahead of its previous target.
>
> Flexible fuel vehicles (FFVs) can run on any blend of gasoline or ethanol.
>
> **"India Targets Alternate Fuels, EVs In Clean Transport Push." Car and Bike, October 6, 2021.**

Although solar and wind power costs are expected to remain higher in 2022 and 2023 then pre-pandemic levels due to general elevated commodity and freight prices, their competitiveness actually improves due to much sharper increases in gas and coal prices, says the International Energy Agency (IEA).

3. Renewable Energy Is Healthier

According to the World Health Organization (WHO), about 99 percent of people in the world breathe air that exceeds air quality limits and threatens their health, and more than 13 million deaths around the world each year are due to avoidable environmental causes, including air pollution.

The unhealthy levels of fine particulate matter and nitrogen dioxide originate mainly from the burning of fossil fuels. In 2018, air pollution from fossil fuels caused $2.9 trillion in health and economic costs, about $8 billion a day.

Switching to clean sources of energy, such as wind and solar, thus helps address not only climate change but also air pollution and health.

4. Renewable Energy Creates Jobs

Every dollar of investment in renewables creates three times more jobs than in the fossil fuel industry. The IEA estimates that the transition towards net-zero emissions will lead to an overall increase in energy sector jobs: while about 5 million jobs in fossil fuel production could be lost by 2030, an estimated 14 million new jobs would be created in clean energy, resulting in a net gain of 9 million jobs.

In addition, energy-related industries would require a further 16 million workers, for instance to take on new roles in manufacturing of electric vehicles and hyper-efficient appliances or in innovative technologies such as hydrogen. This means that a total of more than 30 million jobs could be created in clean energy, efficiency, and low-emissions technologies by 2030.

Ensuring a just transition, placing the needs and rights of people at the heart of the energy transition, will be paramount to make sure no one is left behind.

5. Renewable Energy Makes Economic Sense

About $7 trillion was spent on subsidizing the fossil fuel industry in 2022, including through explicit subsidies, tax breaks, and health and environmental damages that were not priced into the cost of fossil fuels.

In comparison, about $4 trillion a year needs to be invested in renewable energy until 2030—including investments in technology and infrastructure—to allow us to reach net-zero emissions by 2050.

The upfront cost can be daunting for many countries with limited resources, and many will need financial and technical support to make the transition. But investments in renewable energy will pay off. The reduction of pollution and climate impacts alone could save the world up to $4.2 trillion per year by 2030.

Moreover, efficient, reliable renewable technologies can create a system less prone to market shocks and improve resilience and energy security by diversifying power supply options.

Learn more about how many communities and countries are realizing the economic, societal, and environmental benefits of renewable energy.

Viewpoint 4

> *"Cool infrastructure options are increasingly seen as strategies to address extreme heat."*

Cities Can Implement Cooling Strategies to Reduce the Need for Energy

Eric Mackres, Gorka Zubicaray, and Bina Shetty

In this viewpoint, the authors look at ways to design cities that will make them cooler, thus lessening the need for energy to cool them. Through making adaptations in urban design that work with a city's climate, it is possible to reduce cities' dependence on fossil fuels. Eric Mackres is a senior manager at World Resources Institute's Ross Center for Sustainable Cities. Gorka Zubicaray is a senior technical specialist in urban development at World Resources Institute Mexico. Bina Shetty is program head of spatial data and data analytics at World Resources Institute India's Sustainable Cities and Transport program.

As you read, consider the following questions:

1. Who is most at risk from increasing heat?
2. How does vegetation help cool a city?
3. Why is solid data important for cities who want to implement these cooling strategies?

"As the Earth Gets Hotter, Can Our Cities Get Cooler?", by Eric Mackres, Gorka Zubicaray, and Bina Shetty, World Resources Institute, August 10, 2023. https://www.wri.org/insights/earth-gets- hotter-cool-cities-infrastructure-solutions.

This summer the Northern Hemisphere has been so hot with record temperatures—including at sea — that discussions have turned to the limits of human survival. Even in the Antarctic, sea ice is failing to re-form, a drastic departure from normal winter patterns. It isn't just your perception that extreme heat is happening more and more. As a result of climate change, the number of extreme heat events has accelerated—and it's expected to worsen.

Indeed, heat is the deadliest disaster most years, killing an average of 490,000 people globally and causing severe health problems for many more. Deaths from heat are expected to grow by 50 percent by 2050, according to the World Health Organization. But the impact of heat on health isn't equitably distributed—around the world or within our communities. Already vulnerable populations are at the greatest risk.

At the global scale, people in developing countries, particularly in South Asia, Africa, and East Asia—who have contributed the least to cause climate change and do not have the resources to adapt are expected to have their health impacted the most by climate change-induced extreme heat.

At the city scale, neighborhoods with poorer and more marginalized populations, or with worse infrastructure and services, such as fewer green spaces and histories of restricted housing investment, are measurably hotter. This difference within cities exists in part because these neighborhoods are less likely to have tree cover and vegetation, an important mitigant of heat.

These neighborhoods are also more likely to have hard, dark surfaces, which absorb heat.

This is an example of the urban heat island effect, where cities, or parts of cities, experience more heat than rural areas because man-made infrastructure such as buildings, streets, and sidewalks often retain more heat than natural surfaces.

Extreme heat can be devastating, but there are tools every community can use to make measurable differences to reduce heat hazards to health, energy systems, and our economies; improve urban equity; and even curb climate change. By adjusting the same land cover components that are the largest contributors to an urban heat island—such as buildings, trees, and streets and built materials, including concrete, asphalt, permeable pavements, paints, and coatings—cities can dramatically lower their temperatures.

The Influence of Urban Infrastructure on Heat —Examples from Monterrey and Mumbai

In cities, land cover is determined by infrastructure—both gray (roofs, pavements) and green (urban forests, street trees, streams, and reservoirs). And choices among infrastructure options that provide the same function, like stormwater management, can have very different impacts on land cover (for example, vegetated streams instead of paved drainage channels). The relationship between land cover and heat is a consistent finding in cities and research where WRI has worked to identify cooling solutions. And the absence of cooling land cover is often found in neighborhoods where vulnerable people concentrate—exacerbating inequality in exposure to extreme heat.

In Mumbai, where the city government considered heat hazards while developing its first Climate Action Plan (which defines the government's commitment to address climate change), WRI India's analysis found a strong relationship between the share of vegetation cover in city wards and lower land surface temperatures, with a difference of 5.5 degrees Celsius (10 degrees Fahrenheit) between the mean land surface temperature of the hottest and coolest neighborhoods.

Cooler and greener neighborhoods typically have a greater share of high-income residents, while hotter neighborhoods are more often informal settlements. In addition, neighborhoods with a larger share of metal roofs, a roofing material associated with informal settlements and homes of low-income people, often had higher average surface temperatures. Around 37 percent of Mumbai households live under metal roofs and are exposed to higher heat risk.

In Monterrey, WRI Mexico found that the relationship between greater vegetation cover and lower land surface temperatures had very high statistical confidence in 22 out of 27 districts of the municipality.

The findings also show that land surface temperatures vary greatly between districts, with an 11 degrees Celsius (20 degrees Fahrenheit) range of temperatures. The range is still more than 6 degrees Celsius (11 degrees Fahrenheit) even if only considering districts that are mostly urban.

Variations on the pattern repeats itself in cities around the world. Infrastructure, particularly vegetation and built surface types, are critical contributors to the accumulation of heat, how it is experienced by city residents and which residents experience the worst effects.

City Infrastructure as a Cooling Solution

But just as urban infrastructure choices have created areas ripe for extreme heat, the same choices can create neighborhoods and whole cities that are cooler.

Cool infrastructure, both natural and built, can reduce city air temperatures by 3 degrees to 4 degrees Celsius (5 degrees to 7 degrees Fahrenheit). Vegetation, particularly trees, cools through evapotranspiration (releasing water into the air) and providing shade. Solar-reflective built infrastructure, most notably solar-reflective materials used on roofs, streets, walls and other built surfaces, send heat back into the atmosphere rather than letting it accumulate at ground level.

Cool infrastructure options are increasingly seen as strategies to address extreme heat. In particular, a growing number of cities are strategically investing in trees, green corridors and other nature-based solutions, as well as solar-reflective roofs to help reduce the urban heat island effect and the impacts of extreme heat.

In Kochi, India, the city has implemented a tree planting campaign to reduce heat in vulnerable neighborhoods, informed by community knowledge and geospatial data.

In Medellin, Colombia, the city has planted over 8,000 trees to create an interconnected network of green spaces across the city to address heat while improved access to nature and improve biodiversity. City officials estimate that after three years of implementation, the urban heat island effect in the Medellin has been decreased by 2 degrees Celsius (3.6 degrees Fahrenheit)

In Ahmedabad, India, the city together with non-government organizations have developed climate adaptation solutions, including painting white the roofs of 17,000 homes to reduce heat accumulation, to support women living in slum communities.

These cool infrastructure changes provide a myriad of benefits with very low costs, use technologies that are already available around the world, and do not exacerbate climate change (as distinct from mechanized cooling interventions, like air conditioning).

They also provide many cobenefits to help address climate change. By increasing the share of solar radiation that is reflected into the atmosphere and reducing energy demand required for space cooling, cities can ultimately reduce their greenhouse gas emissions.

Adaptation gains are provided by reducing local temperatures, thereby decreasing heat stress, heat stroke and other heat-related health conditions.

Many other economic, equity, and environmental benefits of cool infrastructure have been documented, including reduced energy consumption and peak electricity demand, improved worker productivity, more equitable access to green spaces, improved physical and mental health, and improved air and water quality.

From Information to Action on Urban Cooling

Cool infrastructure solutions have been piloted in hundreds of cities around the world, but there are thousands of cities that can and should adopt these tactics at large scale. However, it's difficult for communities to plan, fund, deploy and track these solutions. A key barrier is actionable data.

Cities and businesses are seeking ways to set targets, prioritize investments and meaningfully measure progress. Without these tools, adoption of projects and policies will remain too slow to save lives. Absent data, investments can't be selected and sited to maximize cost-effectiveness . Changes in urban surfaces are not being measured with methods that are repeatable, scalable and broadly accepted, stifling finance of these solutions, which relies in part on these metrics.

Efforts like the Smart Surfaces Coalition, Arsht-Rockefeller Resilience Center, Cool Cities Network, WRI projects including Data for Cool Cities and Cities4Forests, and others are aiming to address this need by generating local data on heat risk, getting it into the hands of policymakers and informing them about the impacts of their infrastructure choices. The hope is that new data used to power analytical tools that meet the needs of decision-makers can accelerate the adoption of and funding for cool infrastructure to help residents adapt to more extreme heat while bringing emissions down in time to avert even hotter temperatures.

Viewpoint 5

"Renewable energy systems are location-specific: The best system depends on a location's resources (is it windy?), its temporal pattern (how often is it not windy?) and availability of complementary sources (is there hydropower for backup?)."

An All-Renewable Electric Grid Is Possible

David Timmons

In this viewpoint, David Timmons explains the research that exists on the viability of renewable energy as the power source for an electric grid. He argues that it is possible for electric grids to be entirely powered by renewable energy, and that this conclusion is based on research conducted in a number of different types of climates. While adjustments must be made to the types of renewable energy available in different regions and climates, it is possible to accomplish the shift to renewable energy through the use of diverse renewable sources. Timmons argues that the technology exists to make the transition to renewable energy at a relatively affordable cost, but government policies need to change to facilitate this. David Timmons is an ecological economist and an associate professor of economics at the University of Massachusetts, Boston.

As you read, consider the following questions:

1. Why does Timmons claim that islands like Mauritius are attractive for renewable energy transitions?
2. Which two types of renewable energy does Timmons claim are particularly complementary?
3. What environmental impacts do biomass power and hydropower have on the environment, according to this viewpoint? Does Timmons believe these impacts can be addressed?

The main solution to climate change is well known—stop burning fossil fuels. How to do this is more complicated, but as a scholar who does energy modeling, I and others see the outlines of a post-fossil-fuel future: We make electricity with renewable sources and electrify almost everything.

That means running vehicles and trains on electricity, heating buildings with electric heat pumps, electrifying industrial applications such as steel production and using renewable electricity to make hydrogen (similar to natural gas) for other requirements. So the focus is on powering the electric grid with renewable sources.

There is debate, though, about whether fully renewable electricity systems are feasible and how quickly the transition can be made. Here I argue that feasibility is clear, so only the transition question is relevant.

Known Technologies

My research focuses on the economics of renewable energy. To demonstrate feasibility and estimate cost of renewable electricity systems, researchers use computer models that calculate potential production from different technologies at each point in time, based on changing weather conditions. A model reveals which combination of electricity sources and energy storage systems has the lowest cost while always meeting demand.

Many studies demonstrate that fully renewable electric grids are feasible in the United States, Europe, Australia, and elsewhere. My colleagues and I recently completed a small-scale study on the island nation of Mauritius. Islands are attractive places for initial renewable transitions because of their small scale, relative simplicity, and dependence on imported fuels.

There are a number of ways to make renewable electricity: hydro, wind, solar photovoltaics, geothermal, and burning various forms of biomass (plant matter), besides improving efficiency to use less energy. These are mature technologies with known costs.

Other possibilities include wave, tidal, and concentrating solar power, where reflectors focus solar rays to produce power. While these may be used in the future, the need to address climate change is urgent, and in my estimation, the mature technologies suffice.

Opinions on nuclear energy run strong, which is another conversation. But models show that the United States does not need nuclear energy to retire fossil fuels.

The Grid of the Future

Renewable energy systems are location-specific: The best system depends on a location's resources (is it windy?), its temporal pattern (how often is it not windy?), and availability of complementary sources (is there hydropower for backup?). Despite this location sensitivity, studies in disparate places are finding similar results.

Having a diversity of renewable sources can reduce costs. In particular, solar and wind are complementary if the sunny season is not the windy season; models find that a combination of both is typically less expensive than either alone.

For most technologies, larger scale reduces cost. For example, in the United States, large-scale solar farms can be more than 1,000 times larger than residential rooftop systems and about half the cost. To minimize cost, we build large systems.

Because solar and wind conditions vary across the landscape, system costs fall as a production area grows, so there needs to be a robust electric grid to move electricity from places where there is supply to places of demand. We also need more electricity for applications like transportation that currently use fossil fuels. This means the grid must grow.

Studies show that running an electric grid with variable renewable energy will include not using, or dumping, some energy at times, a strategy that reduces cost compared to always storing surplus energy.

Still, some form of electricity storage is needed. Batteries work well for smoothing short-term fluctuations, but for storing energy for many hours or days, pumped hydroelectric storage is less expensive. Pumped hydro uses any extra energy in the grid to pump water uphill, and when energy is needed, the water runs back down to generate power in a turbine. The United States has some existing examples and many feasible locations. With grid expansion, storage may be located at a distance from users.

Hydroelectricity and biomass power are available on demand, so having these in a renewable electric grid shrinks the energy storage need and reduces cost. Both have environmental effects that must be managed.

Hydropower can alter local ecosystems. Burning biomass emits carbon dioxide, but a study I worked on shows that biomass emissions are reversible and are clearly carbon-preferable to fossil-fuel emissions. Sustainability also depends critically on management of biomass fields and forests; the human track record on this has not been stellar.

Renewable energy systems require land. A U.S. study shows that supplying all electricity from wind, water, and solar would need 0.42 percent of land area, plus 1.6 perccent of land area for space between wind turbines. Biomass energy requires much more land than wind or solar, so biomass must be a small part of the renewable energy solution.

Real Barriers Are Political and Cultural

A future renewable electricity grid with associated electrification may or may not reduce energy costs. But avoiding the worst effects of climate change means quitting fossil fuels, whether or not this saves money. Still, the renewable transition will be faster and politically easier if it is less expensive.

In Mauritius, our study finds renewable electricity costs to be similar to present costs there, based on current capital costs for renewable energy. Some studies also find costs for future renewable electricity to be lower than present fossil-fuel costs, in the likely event that costs fall as we build more renewable energy systems and get better at doing it.

And that's it, from a technical perspective. A combination of renewable sources and energy storage—the specific combination depending on local conditions and preferences—can supply all the electricity needed at an affordable price, and will reduce air pollution to boot.

But government policies are needed to make a transition to renewable energy. Climate change is an external cost—borne by society rather than by energy producers—so market forces alone will not make the transition. Besides putting a price on carbon (perhaps with dividends returned to the public), government could make it easier to build the needed infrastructure. And public support is needed: For example, public acceptance of transmission lines to move electricity from the windy Great Plains to city centers is another challenge for an all-renewable grid.

A project on the scale of transforming the energy system will create jobs—many jobs—which is perhaps the economic measure of most importance to the citizenry.

Research from me and others shows that fully renewable electric grids are feasible with current technology at current prices; barriers to using renewable electricity are more political and cultural than technological or economic.

Periodical and Internet Sources Bibliography

The following articles have been selected to supplement the diverse views presented in this chapter.

Casey Crownheart, "Yes, We Have Enough Materials to Power the World with Renewable Energy," *MIT Technology Review*, January 31, 2023. https://www.technologyreview.com/2023/01/31/1067444/we-have-enough-materials-to-power-world-with-renewables/.

Cara Fogler, "Hydrogen: Future of Clean Energy or a False Solution?" Sierra Club, January 4, 2022. https://www.sierraclub.org/articles/2022/01/hydrogen-future-clean-energy-or-false-solution.

Jason Hickel, "The Limits of Clean Energy," *Foreign Policy*, September 6, 2019. https://foreignpolicy.com/2019/09/06/the-path-to-clean-energy-will-be-very-dirty-climate-change-renewables/.

Charles Kenny, "If Everyone Gets Electricity, Can the Planet Survive?" the *Atlantic*, September 29, 2015. https://www.theatlantic.com/international/archive/2015/09/energy-access-sdgs-un-climate-change/407734/.

Karin Kirk, "We've Been Having the Wrong Debate about Nuclear Energy," Yale Climate Connections, July 28, 2020. https://yaleclimateconnections.org/2020/07/weve-been-having-the-wrong-debate-about-nuclear-energy/.

Lingshan Li, "Green Spaces Really Do Make Cities Cooler. Here's How," *Rewilding*, July 18, 2022. https://www.rewildingmag.com/green-space-cool-cities-heat-waves/.

Shannon Osaka and Hailey Haymond, "Buying Renewable Energy Doesn't Mean What You Think," *Washington Post*, June 21, 2023. https://www.washingtonpost.com/climate-environment/2023/06/21/renewable-energy-credits-certificates-greenwashing/.

Jonathan Rauch, "The Real Obstacle to Nuclear Power," the *Atlantic*, February 7, 2023. https://www.theatlantic.com/magazine/archive/2023/03/climate-change-nuclear-power-safety-radioactive-waste/672776/.

Graham Readfearn, "Nationals Say Renewables Can't Meet Energy Needs Without Fossil Support. But Do the Claims Firm Up?," the

Guardian, September 28, 2022. https://www.theguardian.com/environment/2022/sep/29/nationals-say-renewables-cant-meet-energy-needs-without-fossil-support-but-do-the-claims-firm-up.

John M. Reilly, "How Sustainable Energy Becomes Unsustainable," MIT Sloan School, May 3, 2021. https://mitsloan.mit.edu/experts/how-sustainable-energy-becomes-unsustainable.

Weizhen Tan, "What 'Transition'? Renewable Energy Is Growing, but Overall Energy Demand Is Growing Faster," CNBC, November 3, 2021. https://www.cnbc.com/2021/11/04/gap-between-renewable-energy-and-power-demand-oil-gas-coal.html.

For Further Discussion

Chapter 1

1. In this chapter, multiple viewpoints indicate that it is difficult to get policymakers and government officials to support renewable energy policies. Based on what you've read in this chapter, why might that be the case?
2. Based on the viewpoints in this chapter, how can the energy policies of countries like India and China impact other countries?
3. How can the transition to renewable energy benefit the economy, based on the viewpoints in this chapter?

Chapter 2

1. The news about climate change can be distressing. However, the authors of the first viewpoint in this chapter—Robert Lempert and Elisabeth Gilmore—are hopeful that renewable energy can make a difference. From what you read in other viewpoints in this book, are you optimistic as well? Why or why not?
2. Some authors in this chapter argue that the shift to renewable energy sources is not enough. What else do these authors suggest should be done to reduce the impacts of climate change?
3. Based on the viewpoints in this chapter, do you think individual choices can have an impact on climate change and the transition to renewable energy? Explain your answer.

Chapter 3

1. Fossil fuels are a limited resource—when we run out of them, they're gone. Not so with renewables. But can you think of ways so-called renewable energy sources might be renewable, but not sustainable?

2. A survey conducted by the Pew Research Center discussed in this chapter found that while a large majority of Americans support efforts to address climate change, not so many are willing to make changes to their own lifestyles. Does this seem contradictory to you? Why or why not?
3. The final viewpoint in this chapter profiled several young African climate activists. In what ways are these activists trying to prevent climate change? What does this indicate about ways in which people outside politics—especially young people—can impact climate policy?

Chapter 4

1. One of the issues discussed in this chapter is how to get companies to invest in renewable energy when they aren't sure they'll recoup their investment. After all, no matter how much we all want to transition away from fossil fuels, we have to be able to afford it. What solutions are offered in this chapter to address this issue? Do you think they are convincing? Why or why not?
2. Cooling takes a lot of energy, but according to the viewpoint by Eric Mackres, Gorka Zubicaray, and Bina Shetty, there are ways to naturally keep cities naturally cooler so they will use less energy for cooling. What sorts of things can cities or citizens do to reduce the need for energy, so that renewables have a fighting chance to meet the demand?
3. Throughout this volume, authors have disagreed on details, but mostly agreed that if we're to prevent the worst effects of climate change, we must stop depending on fossil fuels. Based on what you've read in this volume, why do you think it is proving so difficult for the world, and especially the United States and Canada, to give up fossil fuels? What can be done to help facilitate this transition?

Organizations to Contact

The editors have compiled the following list of organizations concerned with the issues debated in this book. The descriptions are derived from materials provided by the organizations. All have publications or information available for interested readers. The list was compiled on the date of publication of the present volume; the information provided here may change. Be aware that many organizations take several weeks or longer to respond to inquiries, so allow as much time as possible.

350.org

20 Jay Street
Suite 732
Brooklyn, NY 11201
email: Natalia@350.org
website: https://350.org

As a non-profit organization, 350.org is focused on building a global climate movement through grassroots organizing and mass public actions. It aims to reduce the use of fossil fuels and increase the use of renewable energy sources.

Alliance for Climate Education

4696 Broadway
Suite 2
Boulder, CO 80304
(702) 526-2231
email: hello@acespace.org
website: https://acespace.org

The Alliance for Climate Education is an organization that educates students about climate change and empowers them to lead on climate solutions. It has a youth action network that allows young people to join together and advocate for climate justice.

American Council on Renewable Energy

1150 Connecticut Ave, NW, Suite 401
Washington, DC 20036
website: https://acore.org

The American Council on Renewable Energy is a non-profit organization that unites finance, policy, and technology to accelerate the transition to a renewable energy economy. It works to help build up and finance the renewable energy sector.

Center for Sustainable Energy

3980 Sherman Street, Suite 170
San Diego, CA 92110
(858) 244-1177
website: https://energycenter.org

An independent non-profit organization, the Center for Sustainable Energy is working to use software-enabled program design, administration, and services to transform markets for clean transportation and distributed energy. It works with governments, utilities, and the private sector to accelerate the adoption of clean energy.

Climate Reality Project

555 11th Street NW
Suite 601
Washington, DC 20004
website: www.climaterealityproject.org

The Climate Reality Project recruits, trains, and mobilizes people to take climate action. It was founded by former U.S. Vice President Al Gore and works to encourage urgent global action in support of climate justice and clean energy.

David Suzuki Foundation

219-2211 West 4th Ave.
Vancouver, BC V6K 4S2
(604) 732-4228
Website: www.davidsuzuki.org

The David Suzuki Foundation is an organization that collaborates with Canadians from all walks of life—including government and business—to preserve the environment and create a sustainable Canada. It offers suggestions for how individuals can take climate action online and within their local communities.

Extinction Rebellion

website: https://rebellion.global

Extinction Rebellion is an international, de-centralized, non-partisan movement. Members use non-violent actions and civil disobedience to pressure governments to act with justice on climate and environmental issues.

Green America

1612 K Street NW, Suite 1,000
Washington, DC 20006
(800) 584-7336
website: https://greenamerica.org

Green America uses the economic power of consumers, investors, businesses, and the marketplace to help create a more socially just and environmentally sustainable society. It connects climate issues to finance, food systems, labor, and social justice issues to support a healthy planet and society.

Union of Concerned Scientists

Two Brattle Square
Cambridge, MA 02138-3780
(617) 547-5552
website: https://ucsusa.org

The Union of Concerned Scientists was founded in 1969 to advocate for scientific research directed toward solving social and environmental problems. This organization is dedicated to sharing information, seeking truth, and letting scientific findings guide their conclusions.

Zero Hour

email: info@thisiszerohour.org
website: https://thisiszerohour.org

Zero Hour is a youth-led movement offering training and resources for a new generation of climate-change activists who want to take concrete action against climate change. Their goal is to protect the rights of the next generation to a clean, safe, and healthy environment.

Bibliography of Books

Alice Bell. *Our Biggest Experiment: An Epic History of the Climate Crisis*. New York, NY: Counterpoint, 2021.

Javier Blas and Jack Farchy. *The World for Sale: Money, Power, and the Traders Who Barter the Earth's Resources*. New York, NY: Oxford University Press, 2021.

Bill Gates. *How to Avoid a Climate Disaster: The Solutions We Have and the Breakthroughs We Need*. New York, NY: Knopf, 2021.

Robyn Hardyman. *Biomass Energy*. Minneapolis, MN: Lerner, 2022.

Robyn Hardyman. *Ending the Energy Crisis: It's Time to Take Eco Action!* Minneapolis, MN: Lerner, 2023.

Avery Elizabeth Hurt. *A Global Threat: The Emergence of Climate Change Science*. New York, NY: Cavendish Square, 2017.

Nick Jelley. *Renewable Energy: A Very Short Introduction*. Oxford, UK: Oxford University Press, 2020.

Ayana Elizabeth Johnson and Katharine K. Wilkinson, eds. *All We Can Save: Truth, Courage, and Solutions for the Climate Crisis*. New York, NY: One World, 2021.

Elizabeth Kolbert. *Field Notes from a Catastrophe: Man, Nature, and Climate Change*. New York, NY: Bloomsbury, 2015.

Nathaniel Rich. *Losing Earth: A Recent History*. New York, NY: Farrar, Straus, and Giroux, 2019.

Stephanie Sammartino McPherson. *Hothouse Earth: The Climate Crisis and the Importance of Carbon Neutrality*. Minneapolis, MN: Lerner, 2021.

Henry Sanderson. *Volt Rush: The Winners and Losers in the Race to Go Green*. London, UK: Oneworld, 2022.

Douglas W. Tallamy. *Nature's Best Hope: A New Approach to Conservation That Starts in Your Yard.* Portland, OR: Timber Press, 2020.

Greta Thunberg. *The Climate Book: The Facts and the Solutions.* New York, NY: Penguin, 2023.

Daniel Yergin. *The New Map: Energy, Climate, and the Clash of Nations.* New York, NY: Penguin, 2020.

Index

A

B

C

D

E

F

G

H

I

K

L

M

N

O

P

R

S

T

U

V

W